D1127849

NO REGRETS

I've Enjoyed a Great Life

William F. Scanlin

AuthorHouse™
1663 Liberty Drive
Bloomington, IN 47403
www.authorhouse.com
Phone: 1-800-839-8640

First published by AuthorHouse 8/19/2010

ISBN: 978-1-4520-5252-6 (sc)
ISBN: 978-1-4520-5240-3 (e)

Printed in the United States of America

This book is printed on acid-free paper.

To my wife Dori, my children Sue, Linda, and Bill, and to my grandchildren Jackie, Matt, Tim, Chris, Robbie, and Kenny

PREFACE

WHY DID I DECIDE TO write this book? First, it is intended to serve as a documentation of my life and experiences for my family. Second, I believe that it could be of interest to some of the general public because of the interesting and challenging career I was able to enjoy. At the same time that I was involved in my demanding job, I was able to engage in hobbies and adventurous activities that tended to go beyond those pursued by most people involved in similar careers and financial situations.

Most importantly, however, I am writing the book to acknowledge the key role my wonderful wife played in supporting my work and my hobbies. All the while, she did an outstanding job in raising our three beautiful accomplished children.

My family has been blessed with good health and a love for each other that has sustained us throughout our lives.

CHAPTER 1

I WAS BORN IN PHILADELPHIA on April 16, 1932. My parents were William Fox Scanlin and Jewel Oldach Scanlin. As a very young child, my parents and I lived in Alden and Drexel Hill, suburbs of Philadelphia. Because of my very young age, my recollection of these two residences is rather minimal. When I was about four or five, we moved to Germantown, within the city limits of Philadelphia. The house we moved into in Germantown was a twin that had been owned by my paternal grandmother and grandfather. They had decided to move into an apartment, which was easier for them to care for.

The Rittenhouse Street house in Germantown was built in the 1920s and was unusual in that it had been upgraded by having automatic oil-fired heat. Most homes of that vintage had coal furnaces, which required shoveling coal from a basement coal bin into the furnace. These systems also required shoveling ash from under the furnace grate into metal cans which then had to be carried out to the street for weekly collection. Ashes were sometimes used in the winter to help make traction for getting the car out of the driveway.

While living on Rittenhouse Street, I attended an older elementary school named Pastorius from Kindergarten to third grade. I don't recall anything particularly memorable about my several years of attendance at that school. Play with other kids was pretty much confined to playing with other children who lived on the same block. I recall having clamp-on roller skates, a scooter, and a small two-wheeled bike.

At Christmas time, my dad went all out in putting together elaborate electric train layouts, which were changed each year. While I still believed in Santa Claus, my dad would put the whole thing together on Christmas Eve after I had gone to bed. This took him hours. He usually didn't get to bed until 2:00 to 3:00 a.m.

Summers were spent at the New Jersey shore in a town named Strathmere. When I was four or five, it was routine that I would take walks in the evening with my grandfather to go inspect the pumps that took well water and put it into a tall storage tank that provided water to the community.

At that time a steam locomotive brought several passenger cars from a switch point in Tuckahoe, New Jersey to Strathmere. Basically, this train brought commuters from Philadelphia to the little towns along the Jersey coast. From Tuckahoe, the remainder of the train would go on to Ocean City. As the train came from Tuckahoe, it would pass through Strathmere on its way to the end of the line in Sea Isle City.

Since my grandfather Scanlin had a considerable interest in anything to do with trains and rolling stock in general, he would take me to the train road bed, which was elevated perhaps five or six feet above the surrounding ground. This meant that the steam locomotive appeared to be very large to a four year old. In addition, it made lots of noise as it clanked and hissed while it passed by, and you could feel the heat from the fire box. Needless to say, each one of these train observations was intimidating to this four year old tot. I think my granddad enjoyed taking me there. Although I remember feeling some pangs of fright, I don't recall objecting to these episodes.

A special treat that my granddad provided for the family was that, since he was a senior person with the Philadelphia Rapid

Transit System, his office was five or six floors up, facing on Broad Street. This location provided excellent viewing of the Philadelphia Mummers Parade, which took place every New Year's Day. Another activity that my granddad took me to involved tours of the so-called car barns. These were buildings, where trolley cars and subway cars were stored and maintained. In addition to the regular passenger trolleys, which were dark green, there were work-cars painted orange that were used to do repairs on tracks, on overhead electric wires, and were used to rescue disabled trolley cars.

The house at the Jersey shore (circa 1935 – 1944) was a conversion from an old boat house, dating from the late 1800s. In the boat house a number of rental row boats had been stored and operated by a Capt. Johnson, a distant cousin of my grandmother Scanlin. When Capt. Johnson died, his wife Lucy Johnson retained ownership of the boathouse and her house next door. Somewhere in the early 1900s, the old boathouse was converted into two two-bedroom apartments. My grandmother rented one of these apartments for a number of years.

These apartments were relatively primitive by today's standards. They had no hot water and no refrigerator. Perishable food storage was in an icebox. Twice a week a delivery truck delivered a block of ice. The lack of hot water was somewhat unpleasant, especially in the beginning and end of the summer season because the water in the community water-system tank was rather cool. The cold water meant that showers were taken quickly, after you finally got the courage to jump under the cold stream in the first place. Another aspect of the shower was that the waste water dropped through a hole in the floor, leading to the ground under the house. This situation wasn't as bad as it may sound since the house stood on five or six foot tall wooden pilings, which were necessary because high tides frequently flooded the ground below the house.

A septic tank, which took sewage from both apartments, was located under the house. I recall that it had a concrete wall, which I suppose extended several feet below ground as well as about two feet above ground, making it visible from the back of the house. As

I remember, it was about eight feet in diameter and was covered by a wooden lid.

The main reason for mentioning the tank is that when extreme high tides occurred, the tank would apparently flood, causing a rather nasty sewer smell to occur. I am sure in retrospect that this system, as well as similar systems servicing other dwellings along the bay released lots of bacteria into the bay water. This was clearly an unsanitary condition; however, standards were different at that time, and somehow we all survived.

During summers at the shore, we fished frequently from rented rowboats. In those days, fishing was very good. In an hour or so one could catch a dozen or more fish. The prevalent species were king fish, flounder, weak fish (so named for their weak mouth), and croaker. Occasionally, we would run into a school of blue fish (related to mackerel). All of the fish we caught, we would clean and eat.

A special treat for a youngster like me was to be taken night fishing off the railroad bridge. The bridge was not really meant to be used by pedestrians. It had no railing, and there were spaces between the railroad ties that were open to the water about twelve to fifteen feet below.

I had to be careful where I stepped because a little guy like me could easily have fallen through the openings. In walking out on the bridge, I could always remember some apprehension over what we would do if a train would approach while we were out there. I'm sure the adults knew the train schedules so that there was little chance that a train would come along.

One episode that occurred while we were out fishing in one of the row boats was that my dad told me to look up. As I did, there was the dirigible Hindenberg on its way to its mooring at the airbase in Lakehurst, New Jersey. I also remember seeing the newspaper headlines describing the tragedy at Lakehurst, with pictures of the burning Zeppelin and the pictures of the burned people who had somehow survived.

As I got older (nine – eleven years), my dad switched his interest from boat fishing in the bay to fishing on the other side of the island in the ocean surf. I suppose that this interest was driven by the fact that the ocean fish were larger, and they put up a strong fight, which he enjoyed.

This was especially true as he began to catch several varieties of sharks. Some of the sharks ranged from thirty to fifty pounds. Bathers at the beach were concerned as they watched him bring in the sharks from a distance of perhaps fifty to a hundred feet beyond the bathers in waist-deep water. The fishing rods used for surf fishing were about seven to eight feet long and were made of bamboo. The line on the conventional Star Drag reels was linen, requiring an occasional washing and drying on a rotating rack.

About age eleven, I became big and strong enough to handle a surf casting rod myself. So I joined my dad in fishing the surf. My success was poor since my cast didn't get far enough out in the surf to where the fish were.

The island where Strathmere is located is only about two blocks wide between the tidal bay on the west side to the ocean side on the east. In the late summer of 1944 a strong hurricane was forecast to hit the region around Strathmere. Although it was late in the summer season, my grandmother Scanlin and my great grandmother Fox were still staying in the old converted boathouse. My dad tried to persuade them to come home to Philadelphia. My grandmother said that she had gone through such predicted storms before with no problems, and this storm would probably be the same. The grandmothers elected to stay.

This storm, however, turned out to be different. The tidal surge of the bay and ocean water turned out to be extreme. The two elderly ladies watched the water come up under the house as it often did. But, rather than stopping at several steps below the level of the floor, it kept on rising. They watched the water come on to the floor and continue to rise. They prepared to climb on to the dining room table to avoid the rising water.

Basically, the entire island was inundated and only homes built on substantial wooden pilings were going to survive. Fortunately for my grandmothers, the water began to recede before it got to the level of the dining room table surface. The old boat house was strong.

Although there was some damage to interior furnishings, the house was unscathed.

The day after the storm the weather was bright and clear. My mom and dad drove down to Strathmere from Philadelphia. They were shocked at what they found. Many of the roads were covered with sand and were nearly impassable. Also, many houses were swept from their foundations and were in most cases a total loss. Houses built on cinder-block foundations were especially hard hit. Those set on taller wooden pilings did best. There was no further argument from the grandmothers. They allowed my dad to evacuate them back to Philadelphia. Sadly, they and my family never returned to the old boat house in Strathmere.

CHAPTER 2

OBVIOUSLY, MY EARLY LIFE WAS influenced by my grandparents and my parents. At this point, I would like to recount some of the information that I know about each of them.

My paternal grandfather was born in Philadelphia in 1885. His father was Charles Scanlin, who married Louisa Muntz. It seems that there must have been some Irish somewhere in his background, but our work in family genealogy has been unable to locate it. Clearly Louisa Muntz was of German ancestry. Louisa was small in stature, being perhaps five feet tall. As a result, as a child, I called her little grandma.

My grandfather, William Robert Scanlin inherited the small-stature trait. He was short, about five feet six inches. He graduated from Northeast Technical High School in Philadelphia. He went to work as a draftsman with the Philadelphia Rapid Transit System, also known as the PRT. He rose rapidly through the ranks and by 1930 became Chief Engineer.

He continued in that position until 1938 when he died prematurely from an infection at age fifty-seven. He was known for his great, friendly, personality and was well-liked by everyone he had contact with. Of all my grandparents, he had the most significant influence on my early life while he was alive.

On my mother's side of the family, my maternal grandfather, Albert Oldach, was an adventurer early in his life. Stories passed down through the family told of his adventures in Texas in the early 1900s. When he settled down, he married my maternal grandmother, Margaret Linder.

Albert and Margaret lived in a single-family home in Haddon Heights, New Jersey. He established a book binding business located in a building in Philadelphia near the waterfront of the Delaware River. This location meant that he had to commute from Haddon Heights to his office and shop daily. He took a bus from Haddon Heights to the Camden, New Jersey, side of the Delaware River. He then boarded a ferry to cross to the Pennsylvania side. At that time there were no bridges across the river. Once across the river, he was able to walk several blocks to his shop.

Albert's book binding business was quite successful. His artistic talent enabled him to produce beautiful hand-tooled gold leaf engravings on the leather jackets of his bound books. Since his bound books required so much labor-intensive effort, his limited production items were basically collectible volumes. A very special bound and engraved book was done for the Pope. I am fortunate to have several of his books in our personal book collection.

As a hobby, my grandfather Oldach loved to garden. The property in Haddon Heights was large enough to support a garden containing lots of vegetable plants. My mother told me that her parents used the vegetables regularly in their meals during the summer months.

After some ten or twelve years, the Haddon Heights to Philadelphia commute became very tiresome. The family moved to Philadelphia. They moved into a home owned by great grandmother Linder. This three-storey brick row house was considered large for that time. On the second floor, there was a so-called music room, where an upright piano stood.

On the first floor, at one end of the living room, there was a built-in pipe organ. As a child, I can remember turning on the switch on the side of the organ which started the electric blower located in the basement. After that, pressing on any of the keys, brought forth wonderful pipe organ sounds. Great grandfather Linder had used this organ for his pleasure and to practice, since he was the organist for a Lutheran Church in the center of Philadelphia.

When the Oldach family moved into great grandmother Linder's home, great grandfather Linder had already passed on. My mother Jewel Oldach, along with her brother Carl Oldach lived there during their high-school years. Jewel left when she married in 1931. In 1938 my grandfather Oldach died due to congestive heart failure, only a few months after grandfather Scanlin had died.

My paternal grandmother was born Ella Fox in 1886. On the Fox side of her family, she claimed that she was a direct descendant of a Quaker named James Fox who had settled in Salem, New Jersey. Supposedly, he was in Salem before William Penn settled in Philadelphia. In our work on the Scanlin family genealogy, we have not been able to verify this early branch of the family.

On the maternal side of my grandmother's family, my great grandmother Fox was born Ida Thompson. The Thompson branch of the family was traceable back to the American Revolution. In fact, a verbal account passed on by great-great-grandmother Thompson was that one of her Thompson ancestors was a drummer boy in the Revolutionary Army. The passed-down story is that he was forced to hide in a pig sty to avoid capture by the British.

My grandmother Scanlin never had a career, as was typical of most women of the era of about one hundred years ago. She was, however, very active in her Lutheran Church. She kept meticulous written records in connection with her church work. I still have some of these notes, which are interesting to read, since she recorded the prices paid for various food items purchased for various social activities. She wound up outliving my dad by about a month. She passed on in 1976.

My maternal grandmother Margaret Linder Oldach was born Margaret Linder in 1879. She was the oldest of the three Linder children. Her brothers, Edward and Emanuel became successful businessmen. Edward was an optician with a high-end shop located on Chestnut Street in Philadelphia. As I remember, it was about on the corner of 17th and Chestnut Streets. Emanuel became a candy maker famous for his very fine coated chocolates. His store and his factory were located on Lehigh Avenue in Philadelphia.

When Margaret was married to Albert Oldach, they lived in the single house in Haddon Heights. Somewhat after their move to Philadelphia, Margaret developed a severe case of rheumatoid Arthritis, which gradually affected her entire body. She sought treatment for her condition far and wide, even visiting a specialist in Canada. However, at that time, there was no effective means of dealing with her ailment.

As a child visiting her home, I still recall the large green bottle of Asprin, a medication that she used regularly to ease her pain. Eventually, she was confined to a chair, unable to walk or to take care of herself. A nurse needed to be with her at all times. To give the nurse a night off, my mother would take a bus and two trolley car lines to get from our home to hers. As a child, I accompanied my mother on these weekly visits. Eventually, my grandmother required more structured care. She was moved into a nursing home, where she remained until she died in1949.

My father, William Fox Scanlin, was born in Philadelphia on 15 January 1910. He went to elementary school at Pastorius, the same school that I attended for my first few years. He went on to attend and graduate from Germantown High School, the same high school that my wife Dori attended some years later. During the 1920s, radio technology was developing rapidly. My dad developed an interest in radios and built a number of receivers for his own use and enjoyment. From this interest, he developed good wiring techniques, which he used in assembling the elaborate O-gauge train layouts that I had as a boy.

My dad was an excellent musician. He played the clarinet, the saxophone, the coronet, and the piano. He played all these instruments by ear, since he didn't read music. Between the ages of about eighteen and twenty-one, he was band leader of his own small band. His band played for various events and social functions.

After high school, my dad went to the University of Pennsylvania to study civil engineering. After a short while at Penn, he wanted to keep a job he had with the Pennsylvania Railroad, so he switched to taking night classes in engineering at Drexel Institute of Technology. Unfortunately, he never graduated.

My dad worked for a number of construction companies where he did estimating and expediting of the substantial projects they took on. The estimating work was critical because the project estimate had to be low enough to win the contract, but high enough so that the company could show a profit at the project completion.

As I mentioned earlier, my dad was an avid fisherman when we summered at the old boat house in Strathmere. Another hobby of his was photography. Over the years, he owned a number of different cameras, ranging from a 4 X 5 speed graphic to a Contax 35mm. He would develop his own negatives and then print them in his well-equipped dark room, which he had built in the basement. During the 1930s, he printed our own Christmas cards, which usually featured me as a young child.

During the years of World War II, my dad was young enough to be considered for the military draft. However, because the construction work he was involved with involved projects at the Philadelphia Navy Yard, he was deferred. In the 1950s, my dad became active in the Masons. In fact, when I became twenty-one, my dad served as worshipful master as I went through the three degrees leading to my becoming a master mason.

By the time my dad got into his 60s, his health began to decline. He smoked at least a pack a day for most of his adult life, and he did not get enough exercise. He developed some heart problems and a blood condition where he produced too many red blood cells. Finally, he developed cancer of the esophagus. Due to his weak heart, the doctors elected not to operate on the cancer, which ultimately proved fatal. He died on December 27, 1976.

My mother Jewel Oldach Scanlin was born in Haddon Heights, New Jersey on 15 November 1910. She enjoyed her early years in Haddon Heights and had fond memories of the house they lived in and especially the big central entry hall that the house had. She enjoyed watching the vegetable crops grow in the garden that her father planted and attended to each summer. As a teenager, she was recognized as a fast runner and had a coach who worked on sprint starts, using blanks in a pistol to signal a start.

When the family moved to Philadelphia, Jewel attended Simon Gratz High School. After graduation from Gratz, she went to Pierce Business School. Later Jewel used her clerical skills to help her father with the book binding business. Jewel married my dad before she was twenty-one and needed her father's permission to get a marriage license. My mom and dad were married on 10 June 1931. I was born ten months later on 16 April 1932.

My mother was somewhat over-protective of me, especially since I was an only child. Since I was inherently cautious as a child, I never developed athletic skills that I might have been capable of had I had more encouragement at home.

Jewel was an excellent cook and homemaker. She had learned many of her cooking skills from her mother and maternal grandmother. Having been brought up with musically-talented family members, Jewel enjoyed music very much; however, she never learned to play an instrument. In her teens, Jewel learned to drive. She prided herself in being able to drive a 1920s era Buick sedan. This was an accomplishment for a young girl because those cars had no power steering, power brakes, or an automatic transmission. She continued to be a good driver until her vision began to fail in the late 1980s.

After my dad died in 1976, Jewel continued to live alone in the house in West Oak Lane. However, she had no close relatives living in the Philadelphia area. I, therefore, encouraged her to sell the house and move to Livermore, California. She was happy in Livermore, living in a rented senior citizens condominium. She would drive her older lady friends to lunch and dinner at restaurants in the Livermore-Pleasanton area.

Unfortunately Jewel developed macular degeneration and leukemia and gradually lost her sight. She eventually became totally blind and lost her independence since she was no longer able to drive a car. She coped with her blindness very well and continued to live alone, and, by feel, she was able to prepare meals for herself. Jewel's Leukemia eventually proved fatal. Jewel died in 1995 when she was almost 85 years old.

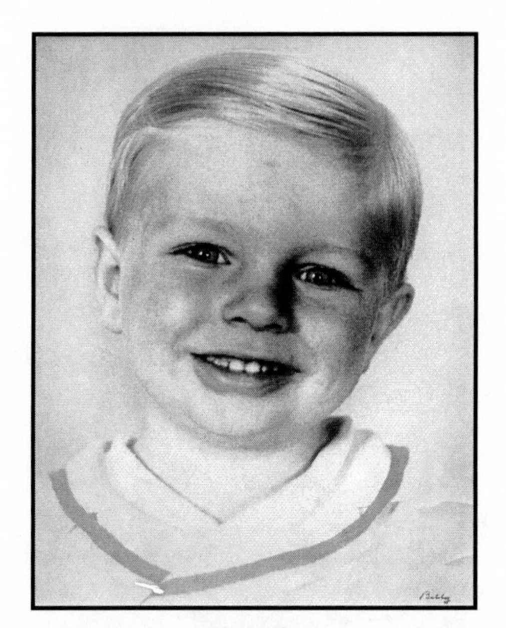

Age 3

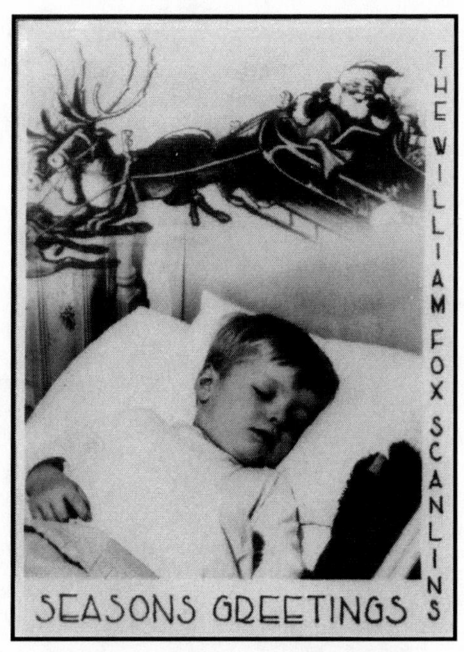

Age 3 – Father printed cards

Age 4 With Paternal Grandfather

Drexel Hill House

Age 12 With Dolphin

Converted Boathouse in Strathmere

Strathmere After Hurricane in 1944

Tulpehocken Street House

CHAPTER 3

IN THE TIME FRAME OF 1940 – 1941, my dad's income had fallen somewhat because of a change of jobs. This resulted in my parents wanting to reduce their mortgage payments. The house on Rittenhouse Street was put up for sale. They found a new smaller row house located just barely within the city limits in West Oak Lane. The new house was a two-storey brick home, with a basement and garage at the ground level. It had three bedrooms and a bathroom on the second floor. Like the house on Rittenhouse Street, the new house had an automatic oil-fired furnace and a gas hot water heater. The way the houses were configured there was a back alley that gave access to the garage at the rear of the house.

The day established to move from Germantown to West Oak Lane was set for March 1941. About a week prior to moving, I contracted the measles. My mother, although very concerned about my illness, decided not to call a doctor. At that time, a doctor was required to quarantine a house having anyone with a contagious disease. The quarantine interval would have made it necessary to change our moving day. My mother read all she could about treating measles and nursed me through, even though for several days I had a very high fever. On moving day, while still recovering, I was bundled up, driven to the new house and immediately put to bed.

In the new house I still recall very clearly the evening of 7 December 1941. Newspaper boys came up and down the streets crying out about the Pearl Harbor attack and selling papers giving more details. World War II had begun. In time, it affected everyone's life. Gasoline was rationed. If you didn't have a reason to use your car for your livelihood, your allocation (an A sticker) was very small. Since my dad was able to use public transportation for getting to and from work, he sold our car, a 1941 Hudson, which we had had for only a short time.

Sugar, meat, and shoes were in short supply and were also rationed. Everyone got ration books which contained stamps which you tore out to obtain the various commodities. Sometimes even though you had enough stamps, certain items were simply not available. Sugar was an item that frequently was not on shelves. When some store got a shipment of sugar, all the housewives would hurry to that store to get some sugar before the supply ran out. Often a store would re-bag sugar into two-pound brown unlabeled paper bags that were limited to one per customer. Desserts were necessarily curtailed in most households.

Shortages of butter meant that margarine was often used as a replacement. Laws at that time didn't allow margarine to be colored yellow. (The dairy industry must have had a strong lobby.) The margarine package, therefore, was sold with a capsule so that at home you could add the coloring to make your margarine yellow, so that it would look like butter.

As the war effort cranked up, I became very interested in all the equipment used by the military. Hobby stores sold kits with which I built models of aircraft, ships, tanks, and artillery pieces. All of the furniture in my bedroom was covered with various models. I was fascinated by all of the data pertaining to the equipment. I knew and, in fact, still remember the horsepower of the engines in the fighters and bombers of not only those of the allies but also those of the Germans and Japanese. Even as a youngster, I had a desire to fly aircraft, a desire I was unable to fulfill until I was twenty-nine. I also knew the characteristics of our warships, their displacement, how many and what size guns they carried. Knowing the displacement of ships comes into play later.

When we moved to West Oak Lane, I was transferred to a different elementary school. The school was named Pennypacker. I entered into the third grade and had a teacher who found I had some shortcomings that had developed while I attended Pastorius. The teacher advised my mother to coach me for a while to get me up to speed. This effort paid off since within a year I was performing at a level commensurate with the top pupils. There was one criticism, however, that was passed on to my mother. One of my teachers was teaching us to read music. As a part of this activity, she went around the room having each child sing a line or two to the class. Because of my shyness, I simply refused. The teacher encouraged perhaps even threatened but to no avail. I would not sing.

One memorable event that took place during the fourth grade was that two other boys and I were selected to be the wise men in a Christmas pageant. One of the other boys was at that time my best buddy. In the pageant, the three of us paraded across the stage while the Three Kings carol was playing, looking up for the star in the east. Interestingly, my friend in that play and I recently reestablished contact with each other after nearly fifty years. He also remembered well our participation in that play so long ago.

In West Oak Lane, my dad continued the elaborate electric train layouts, but because I was now older, he could start planning and putting together the layouts a week or two before Christmas. I was of course very interested in this effort and, in fact, after several years, I took on the entire project myself.

I was very interested in all things mechanical. With the electric train, I would take apart the engines, clean the motor commutator and replace brushes. Power tools, like electric drills got the same disassembly-reassembly treatment. At home one day I observed a service technician service our oil-burning furnace. Having seen what he did, I advised my folks that next year I would do the servicing. This involved brushing the soot from the boiler, removing the oil injection nozzle, cleaning the filter screens and making sure that the small orifice was unobstructed. After this work, I would start up the furnace and adjust the air- fuel ratio until I got a soot-free (non-smoky) flame. I was doing this at about age fourteen.

As I indicated earlier, my dad had many interests beyond music, radio, fishing, and photography. He read books on astronomy, and was familiar with the characteristics of the planets, stars, and constellations. We often took walks in the evening, and he would point out to me stars, planets, and constellations visible in a clear night sky. He became interested in foreign language and spent a considerable amount of time self-educating himself in Spanish. Being familiar as he was with the construction business, he felt that a bright future was ahead in the construction field down in South America. He was also a golfer, and when I became old enough, we went to driving ranges and played golf at various public courses in and around Philadelphia.

Since my grandmother no longer rented the place in Strathmere, New Jersey, in the summer of 1945 my folks were invited to use an apartment owned by my uncle Emanuel in Townsends Inlet, New Jersey. At age thirteen, I made a number of friends there. We enjoyed the beach and the ocean surf, as well as swimming in the bay on the other side of the island. During that summer, World War II was winding down and everyone was looking forward to the surrender of the Japanese.

In August, I was walking past a little market on the main street where we did most of our shopping. Some newspapers were stacked outside and I noticed that we had dropped an atomic bomb on Japan with a yield equivalent to approximately 20,000 tons of TNT. From my interest in such matters, I was astounded that a single airplane could carry a bomb with such force. I knew that some of the largest conventional bombs carried by allied bombers, known as blockbusters were about 8,000 pounds. In this case, the bomb had explosive power of a mass of TNT about the weight of a heavy cruiser. I had no idea at the time that I would spend most of my working life working in the field of atomic weapons.

I recall clearly the day we got the news that the Japanese had surrendered, ending World War II. Everyone in Townsends Inlet was excited, including my mom and dad. All of the local bars that afternoon did a brisk business as everyone celebrated.

Returning to Philadelphia that fall, I was playing a game of touch football in the street in front of our house. As I jumped to catch a pass, I was hit by an older boy. When I came down, I knew right away that my right leg was broken. No one believed me. Even my mother, who came out of the house to look after me, encouraged me to get up on my feet and walk into the house. As I tried to stand, you could hear the ends of bones grinding together, and everyone knew that I had a serious problem.

That evening an ambulance took me to Temple Hospital, where a leg cast, covering my leg from ankle to hip was put on. I spent the next three months in the cast getting around on crutches. Since I was in Junior High School, it was felt that teenage kids, as active as they are, could put my healing leg in jeopardy. I was, therefore, home-schooled, with a special teacher coming to the house several hours each day. This home-school teacher provided little input in Math. When I returned to junior high, I had some difficulty catching up with the algebra that the class had been getting.

My junior high experience ended with the eighth grade. The high school I was to attend had a four-year curriculum. I, therefore, attended Central High School in Philadelphia from 1946 – 1950. Central High School was unusual in respect that it was a public school having a limited number of students. The student body was all male, as was the faculty. Only students with high achievement levels were accepted. If a student's performance did not measure up, they were immediately transferred to another high school. Clearly, everyone who attended was likely to go on to college.

The school had a very long tradition, dating back to 1836. One of the later buildings where I attended even had its own planetarium, a rarity at that time, since the building was built in 1939. Academically, it was so superior that upon graduation the students received a BA degree. In earlier days, this allowed graduates to enter college at the junior-year level. During my years of attendance at this school, I did well, keeping myself on the honor roll.

I graduated from Central in June of 1950. On graduation evening, my family was in the high school auditorium. When the ceremony was over, I went out to retrieve the family car and heard on the radio that North Korea had invaded South Korea, beginning the Korean War. This act underscored the understanding that we were in the midst of a cold war that was to have a profound influence on my eventual work in the field of nuclear weapons.

CHAPTER 4

AS I APPROACHED COLLEGE AGE, my parents offered me a choice of whether to live at home and attend the University of Pennsylvania, or to live away, attending Penn State University. I chose to go to Penn State and began college, living in a brand-new dormitory in a single room. One of my best friends from high school was in the same situation. I was enrolled in mechanical engineering, while my good friend Don was enrolled in aeronautical engineering. During the first two years, the curriculum was almost identical.

When my sophomore year arrived, I was moved to an older, less desirable, dormitory complex. It was located out near the periphery of the campus. No single rooms were available in that complex, so I shared a room with Dick,the son of the minister where I sometimes went to church in Philadelphia. During the summer before my sophomore year, I began dating my future wife Dori. Dori and I had known each other for some time, but for some reason or other we had never seemed to connect. One hot summer morning after church a group of us decided to go swimming at a lake. After that day, neither one of us ever dated anyone else again.

During our sophomore year, Dick and I decided to buy a car. We jointly bought a 1937 Ford for $75.00. Sloppily painted on its trunk were the words, "The Roach." With some brake fluid, this name was quickly removed. Dick and I became very interested in cars. We did repairs on the old Ford as necessary. We even rebuilt the transmission in our dorm room, after breaking a tooth off the cluster gear in demonstrating its wheel-spinning ability. Needless to say, during that year, our grades suffered somewhat.

Dick was a gymnast on the Penn State gymnastic team. In order to compete, he had to work out regularly. I also began a gym work-out regimen, which I have continued throughout my life

Don and I both had fathers in the commercial construction business. Our fathers were able to use their influence to get us summer jobs as ironworkers. We each had to obtain a union card and went to work as apprentice ironworkers. At that time, a new steel plant was under construction at Morrisville, Pennsylvania. Since the steel mill involved much heavy equipment, the foundations and slabs beneath the mill were very thick and contained huge amounts of reinforcing steel.

Our job as iron workers was to carry reinforcing bars, mostly an inch and a quarter in diameter and thirty feet long. The two of us would pick up one of these bars and carry it down into the pit of the foundation. Once having put it into place, we would go back and get another one and repeat the process.

The veteran iron workers were betting that the young college punks would never survive the day. We worked all day long shirtless in ninety degree heat carrying the heavy hot steel. We did survive that summer and became quite strong and tough. We worked so hard and the heat was so intense that even the journeymen would sometimes throw up and have to take a break.

Knowing how tough the work was, in succeeding years I would go to the gym and carry out a conditioning process weeks before summer vacation began. I would lift and carry a heavily-loaded barbell on my shoulder to simulate the effect of carrying the reinforcing steel. The money we earned for this work was very good, especially since we were paid double time for overtime and for work on Saturdays when it occurred. Working shirtless all day in the sun and getting covered with rust and scale from the iron, I would come home each evening looking very dark brown or black.

For my junior year at Penn State, I was in the same dormitory complex as my sophomore year. The minister's son had graduated so I had a new roommate. My new roommate was a World War II vet, taking industrial engineering on the GI Bill. The vets were tough competition academically because they took college more seriously. A number of them were married and lived in a trailer complex on the campus grounds. My grades began to turn around that year as I spent less time working on cars and found the more advanced engineering courses more interesting to me.

During that year, I made friends with Jim, a guy in the same dorm complex. He was a mechanical engineering student and had a strong interest in cars like me. For our senior year, we decided to room together. We sought a room in a private home off campus. The room we found was in a rather old house owned by an elderly couple, located about four or five blocks from campus.

Several other rooms were occupied by students that we didn't know. Since the house had only one bathroom, bathing and toilet needs were sometimes problematic. Otherwise, our accommodations were quite satisfactory. Jim was a couple of years older than I since he had done a tour in the Merchant Marine prior to starting college. He was a stabilizing influence on me I believe. We each had the habit of putting our cars away during the week in the garages that we rented.

Another profound influence on my life, during the later years of my college education, was that I was dating Dori. In fact, we had committed ourselves to marry as soon as I finished school. This commitment to my future wife helped me to knuckle down and finish my studies more strongly as I approached graduation.

The engineering curriculum at that time required a continual heavy course load each year to be able to graduate in four years. Also, I had opted for the Advanced ROTC Program (Air Force) which exacerbated the course-load problem. As a result, as I approached the end of my senior year, I needed three additional units to satisfy graduation requirements.

Due to my fear of public speaking, I had put off a speech course until the very end. It was necessary for me to spend an additional three weeks at the college to complete the speech course during summer school. Dori and I had to put off our wedding day until after this course was completed. Our wedding day was set for 26 June 1954. Although I had much apprehension over the speech course, I did fairly well, obtaining a B+ even though I had cut the final day's class in order to be able to drive to Philadelphia in time to attend the wedding rehearsal and dinner.

As I had indicated earlier, since my childhood, I had always wanted to be a pilot. I had hoped that I would gain acceptance to pilot training in the Air Force. Planning for accepting my commission had gone so far as to getting Air Force uniforms and attending Air Force summer training camp.

That camp, which occurred between my junior and senior year, was about six weeks long and was like basic training. During camp, we did marching drills, learned to fire small arms, trained with gas masks, and learned about weather and navigation. We also got some exposure to aircraft in that we got to have a flight of several hours in a B50 bomber and were even able to explore the insides, including the cockpit, of the gigantic ten-engine B36 bomber.

Since the Korean War was going on throughout this period, it was my expectation that upon graduation I would be accepting a commission of second lieutenant in the Air Force. My physical examination had indicated that I was nearsighted in one eye and was thus unable to qualify for pilot training. As the Korean conflict wound down, the Air Force had decided that it would only accept newly-qualified officers who were able to qualify for pilot training. Any plans that I had made toward receiving an Air Force commission had disappeared.

With the option to going into the Air Force gone, it became necessary for me to start scheduling interviews with companies I might wish to work for upon graduation. Because of my interest in cars, I had scheduled an interview with General Motors, seeking a position in their engine-design department. Although the interview seemed to go reasonably well, I never received an offer. I did receive offers from three other entities. One was from the Atlantic Refining Company in their refining department. Another was from Curtiss Wright and thirdly from Picatinny Arsenal in Dover, New Jersey. Although Atlantic Refining and Curtiss Wright offered a higher salary, my offer from Picatinny was with their Atomic Weapons Laboratory, dealing with the design of atomic weapons.

My final interview with Picatinny was with Robert Schwartz, the Chief of the atomic laboratory. It turned out that he had been a ballistic and mechanical designer of the first U. S. atomic shell, known as the Mark 9. His laboratory was rather small, with interesting work to do, and seemed to offer the possibility of rapid advancement. I decided to accept the offer with Picatinny and planned to go to work as an ordnance engineer.

In order to complete my final interviews and paperwork with Picatinny, Dori and I drove from Philadelphia to Dover, New Jersey, on April 16, 1954, my birthday. It was a cold and rainy day. My acceptance papers were completed, and at the same time, Dori applied for a secretarial position in the research and development laboratory administrative office. Thus, we both came away with new jobs, having a scheduled start date after the completion of our honeymoon in early July of 1954.

Our stay at Picatinny was a little longer than expected. We were in a hurry to get home to a special dinner that my mother had prepared for my birthday. As we approached the front of the house in West Oak Lane, the parking space was tight. As I hastened to get in, I caught the rear fender of our Ford with the front bumper of my dad's car. Since I was moving rather smartly, I virtually tore the fender off the old Ford. There was also some minor damage to my dad's car. After we commiserated a bit about this last unhappy event, we decided that we should get on with the birthday celebration since the day had been so successful in other ways.

Several days later, I returned to Penn State with my 40 Ford, without its right rear fender. Dori and I talked about taking the graduation money my parents had given me, supplemented by additional money from Dori out of savings that she had accumulated, to purchase a new car, rather than fixing the old damaged one. With these funds and a generous trade in, considering my old car's age and condition, I purchased a brand new two-door 1954 Ford V8 sedan. Thus, we were able to start married life with brand-new car.

Following the completion of my classes, Dori and I were married on the 26th of June, 1954. We were married in Faith Evangelical Church in West Oaklane, the church Dori had been associated with all of her life. The wedding was well-attended by lots of family and many friends. My best man was Jim, the college roommate that I had during my last year at Penn State. A large reception followed. We honeymooned on a trip to Washington, D. C., Williamsburg, Virginia, and the Maryland seashore.

We then proceeded to Dover, New Jersey, to occupy a brand new one-bedroom garden apartment. We arrived after dark. Upon entering the apartment, we found that there was no electrical power. The fuse box was empty. We had to make the bed in the dark. We discovered that some of our furniture had been delivered to Dover, Delaware, instead of Dover, New Jersey. A few days later we started work at Picatinny.

Logistically, our situation was very good because the atomic laboratory where I worked was located on the second floor of one of the main buildings, while the office where Dori worked, was located in the same building on the first floor. As I began work, I was asked in what general area I would like to work. Since I felt that I was pretty much a hands-on person at that time, I said that I would like to focus on the testing area.

Outside Penn State Dorm in 1952

1940 Ford

Air Force ROTC Summer Camp 1953

Our Wedding – 1954

CHAPTER 5

MY WORK BEGAN TESTING ARMING components, which were used in the sixteen inch, 280 mm, and eight-inch atomic projectiles. As I performed this testing work, it became clear to me and my superiors that I could make important contributions in the design area.

The actual flight testing of the arming components in inert projectiles was quite expensive. I, therefore, designed a special test projectile (I still recall the designation of that projectile, T5040). This projectile was configured such that it could test four times the number of arming components in a single shot while duplicating the launch and flight environment of the real Mark 33 nuclear projectile. This projectile needed to be ballistically matched to the nuclear one. In addition, all of its interior structure had to survive the extreme launch environment produced by the eight-inch cannon.

Parts for four of the T5040 projectiles were built for flight testing at the Yuma Proving Ground in Yuma, Arizona. Since I wanted to witness these tests and participate in the projectiles' assembly, I took my first trip to the west coast during the winter of 1955. My flight from New York landed in Los Angeles. After an overnight, I caught a local flight from LA, down the coast to San Diego, and from there flew southeast to Yuma. I directed and helped the proving ground technicians assemble the shells, and the following day they were fired, with all the arming components working properly.

Around 1956, concern developed over impending Communist exploits in Lebanon. As a result, the army was asked to develop an emergency nuclear capability for the new, at that time, Little John missile. The Little John was only about eleven or twelve inches in diameter and was light enough that the missile and launcher could easily be transported by helicopter. The Little John was slated to have as its final warhead the W45, but the W45 was still in the R & D stage and would not be available for several years. The interim emergency capability warhead was to make use of major parts of the Mark 33 nuclear projectile.

Because the launch and flight environment of the missile was so different from that of a cannon launch, an entirely new fusing and firing system had to be developed. Collectively, all of the parts necessary to accomplish this change were known as an adaption kit. The adaption kit, plus the appropriate parts of the Mark 33 nuclear projectile would then constitute the interim warhead for the Little John missile.

I was assigned the responsibility for the design, development, and production of the adaption kit scheduled to be completed within the short interval of only eighteen months. I devised a new system of arming components activated by a gas generator. Launch accelerometers opened arming valves only when a proper missile launch had occurred.

I performed all of the design calculations and oversaw the preparation of all drawings and specifications. I oversaw fabrication contractor efforts and coordinated and monitored the redesigned fuses that were produced by Frankford Arsenal in Philadelphia. A number of local tests were conducted and some minor adjustments to the design were made.

Eventually, complete adaption kits were mounted to inert Mark 33 projectiles. They were flown on live flight tests at White Sands Proving Ground. This testing proved successful and a number of adaption kits were available to put into the U. S. warhead inventory. The short eighteen-month time scale for development and production was met.

Due to the program's success and my contributions to it, I was nominated for the Arthur S. Fleming Award for outstanding employees. I had heard later Picatinny management had considered me to be too young (I was twenty-five years old at the time) and withdrew the award submission.

Another nuclear program that I was involved with was the development of the nuclear capability for a 155 mm projectile. On this program I was assigned to be chairman of the so-called Shell Ogive Committee. This committee coordinated the efforts of Picatinny, Frankford Arsenal, Harry Diamond Labs, and Sandia. The nuclear warhead portion of that projectile, called the W48, was designed and developed by the Lawrence Livermore Laboratory. Eventually, that system entered the stockpile and remained there for many years.

As nuclear warheads became smaller and lighter, there was considerable activity in the area of tactical nuclear weapons. This work enabled the army to be able to consider arming the infantry with a short-range, low-yield system (Davy Crockett). Davy Crockett employed two sizes of recoilless rifles producing maximum ranges of two thousand and four thousand meters. I contributed to this program by designing the aft portion of the projectile body which housed the fusing and firing components. The entire Davy Crockett effort, which also included delivery system work by Frankford Arsenal, drew many accolades from senior Army officers and officials.

It was during this interval that I became a GS-13 supervisory engineer, heading up a group of engineers and technicians (numbering about eight). This group was responsible for the mechanical design of the tactical systems under development and the ballistics which determined the flight characteristics of the projectiles and missiles. I had achieved this position at age twenty-six.

One sidelight that occurred during this interval and before was that several of us were encouraged to take graduate-level courses at the Stevens Institute of Technology. These courses emphasized ordinance application of engineering principles, including ballistics and mechanical design. Although I completed several of these courses, I did not receive an advanced degree. Work and family demands had caused me to discontinue the course work.

CHAPTER 6

WE STAYED IN THE GARDEN apartment, located in Rockaway, New Jersey, for a period of about two years. With both of us working, we tried to save as much money as possible so that we could accumulate enough money for a down payment on our own home. We looked around the area and finally located a development in Succasunna, New Jersey, where there were small well-constructed three-bedroom homes that were in our price range.

Surprisingly, most of the homes had no garage, but a split level model did. We opted to buy the three-bedroom split level, having three bedrooms on the upper level, a family and laundry on the lower level, and the living room, dining room and kitchen on the first level. When we bought it, it was still under construction. As a result, we were able to select floor coverings and appliance colors. We needed a refrigerator, washer and dryer. We bought all three at a local appliance store and the cost was a bit over one thousand dollars. For that time, and considering our salaries, that was a lot of money (probably the equivalent of $5,000 to $10,000 in today's dollars.).

Prior to moving in, Dori had become pregnant with our eldest daughter. We moved in December of 1956. The winter was bitterly cold. I recall that our single- paned aluminum framed windows were covered with ice on the inside. As we were getting established in our new home, we needed an outside TV antenna. I contracted with an appliance store to install one on the chimney. We were away from the house one day and came home to find that the antenna had been installed.

As we opened the front door to go inside, the odor of fuel oil was very strong. We quickly went down to the lower level and were horrified to see the asphalt floor tiles floating in fuel oil all over the floor. When the contractor installed the antenna, the installers drove in a grounding stake, connected to the grounding wire. This severed the fuel line going from the buried outside oil storage tank to the furnace in the laundry room. The installers were covered by insurance, but the clean-up process of oil and partially-dissolved tiles was terrible to say the least. The fuel oil odor took a long time to disburse.

As spring approached, we began the process of landscaping our lot. The front yard had been rototilled and seeded with grass seed while the back yard we raked and seeded ourselves. Normally, rain was frequent enough and sufficient that no irrigation system was needed. Nevertheless, we were concerned that our quick and dirty planting of the back yard might not be successful.

One morning we awoke and were amazed by seeing a huge amount of water covering the entire back yard. We noticed that our neighbors did not have puddles in their yards. It turned out that the water line from the street to the house had burst and flooded the yard. The builder fixed the pipe, but as a result we got a fine stand of grass in our back yard.

In addition to the grass, we had planted some shrubs in the front of the house, but needed some additional items for the back yard. At the time, there was a great deal of undeveloped land surrounding our housing development. One Saturday, we took a walk, searching for a small tree that we might want to dig up and plant in our back yard. We found one and took a pick and shovel to the site, dug up the tree, and moved it into the back yard.

It was still early spring so there was no greenery yet on our new tree. It was a warm day and we dug the hole for the root ball. I was sweating under the long-sleeve sweat shirt that I was wearing. Later, as I cleaned up I noticed that I was developing a red rash on both forearms. Apparently, the new tree had had some dormant poison oak vines on its trunk which gave me a case of poison oak that was quite severe. The rash developed into a mass of oozing blisters.

I was unable to bath myself and Dori and I were both concerned that it would spread from me to her, which in her case could have been rather serious since she was only a couple of weeks from her delivery date. Fortunately, none of our concerns materialized. In a couple of weeks, my poison oak began to heal and on the 29th of March, 1957, when our daughter, Susan was born, we were both in good health. As it turned out, Suzie was born on the day that had been predicted. After only five hours labor, we had a healthy eight pound, three ounce, daughter.

After our daughter Suzie was born, I was very busy at work. A special safety review with respect to adaption-kit design was sanctioned. This review was conducted as an adjunct to our normal design and development activities. It was, therefore, carried out as work in the evenings and on weekends. We were paid overtime for the duration of this study. It was timely for us because we were able to use the extra money to complete the furnishing of our new house. Later on, Dori became pregnant again. Her pregnancy was perfectly normal and Dori gave birth to our daughter, Linda, on September 23, 1958.

First New Car – 1954 Ford

First House – Succasunna, N.J. 1956

Nuclear Event – Atomic Canon – 1953)

CHAPTER 7

IN THE SUMMER OF 1960, the word came down that management was looking for a senior individual to represent Picatinny at the Lawrence Livermore Laboratory in a liaison capacity. My initial reaction was to decline, since I had a very good and responsible position already. However, after some further consideration and some encouragement from Dori, I decided to accept. In October 1960, we packed up our two daughters and everything we could cram into our little 1960 Corvair Coupe and proceeded to drive to Livermore, California, for a six-month stay.

The drive was an adventure, as we took in the sights of the Grand Canyon and the Painted Desert. Throughout the trip, we took pictures of our cute little blonde daughters. One of our interim destinations was Las Vegas, where we stopped for a day or two. We all enjoyed sunning and swimming at a hotel swimming pool, regrouping from the long drive we had already made.

From Las Vegas, we made the decision to drive north on Highway 395. This turned out to be a bit of a mistake since an early snowfall had already begun on the eastern slopes of the Sierra Nevada. This caused some difficulty on Highway 395 but was much worse as we proceeded to try to cross the high passes heading to the west.

A permanent-looking sign, saying that chains were required, did not deter me from continuing upward. The Corvair had a rear engine, with good weight over the back wheels, and since I drove all the time in New Jersey without chains, I felt sure that we would make it to the top. As we continued, conditions worsened and, as good as the Corvair may have been, it finally bogged down and was unable to move.

Before long a highway patrol car came down the mountain and the patrolman asked why we did not have chains. I tried to explain, but he said, "I will call a tow truck. See you at the top of the hill." The tow truck towed us to the top of the grade. The highway patrolman had sympathy for us and offered to forgo a fine since we still had to purchase chains. The chains would be expensive since they needed to be delivered from the bottom of the mountain.

We proceeded on, with our newly installed chains. As we came off the mountain, we broke out into bright sunlight as we approached Sacramento. When we arrived in Livermore, we stayed in the Townhouse Motel, located on First Street. We stayed in the motel a week or so while we sought more permanent accommodations.

We were unfamiliar with the area, so we began looking for an apartment in the Hayward-Castro Valley area. Returning to Livermore from the east bay on Highway 50, we drove up behind a car occupied by some old friends who had formerly worked at Picatinny. We stopped and talked with them and they mentioned some places that we might look at in Livermore.

We found a brand new apartment complex located on East Avenue called the Biscayne. It looked great to us, so we signed up for a two-bedroom unit that would house us for our six-month stay. The apartments were fully furnished, so getting established was quite easy. Our daughters Suzie and Linda were not yet of school age, so they were able to spend the day using the complex pool and playing with other young children who occupied some of the other units.

CHAPTER 8

AS I BEGAN MY WORK at the Livermore Laboratory, I was assigned to work in a division which had the responsibility of the physics design of nuclear-weapon primaries. The division was known as B Division. A similar physics design division called A Division was responsible for the design of thermonuclear systems. Within B Division, I was assigned to a group looking at advanced concepts, with an emphasis on tactical systems, which were items of particular interest to my home organization at Picatinny.

Our group explored concepts for different warhead designs that would provide greater yield and efficiency for certain nuclear artillery projectiles and missile systems. The work was challenging and interesting and I was able to contribute to the effort by influencing the designs in ways that would assure that the desired flight characteristics would be achieved.

We also investigated using new concepts for atomic demolition munitions, which would allow a building-block technique to be able to achieve the desired yields. As a part of this effort, different physics configurations were to be utilized which would reduce the radioactive fallout from the explosions occurring on the surface.

My B Division group leader had his private-pilot's license, and he belonged to a flying club, which at that time was mostly for the benefit of flyers from the Laboratory. It was called the Flying Particles. With some encouragement from my group leader, I joined the club and began to take flying lessons. The plane that the club had was quite small and its engine only had 65 hp. It had no electrical system. To start it, you had to rotate the propeller by hand.

While I was learning, I would sometimes go out to the airport before work, early in the morning, start the engine myself, run around, jump in the cockpit, and take off. I flew frequently, therefore, between January and April I was able to accumulate the hours and the skill so that by April of 1961 I passed my private pilot flight exam.

CHAPTER 9

SINCE DORI AND I CONSIDERED that our stay in California would only be six months, we took every opportunity we could to explore the sights and places in the bay area, as well as, on down to southern California. Although our daughters were quite young at the time, they still have recollections of our Christmas holiday trip to Disneyland.

In general, our six-month interlude at the Livermore Laboratory had been a pleasant one. However, we began to look forward to our return to our little house in New Jersey. Toward the end of my stay, my group leader and I traveled back east to visit Picatinny to report on some ideas that we had been working on. It was during this trip that my group leader asked whether or not I had given any thought to coming back to work at the Livermore Laboratory full time. I admitted that I had not, but I agreed to give the matter some thought.

Since Dori and I are both only children, we realized that a major relocation of our family three thousand miles from our parents would be somewhat traumatic for them. Upon returning to Livermore, Dori and I discussed the potential move and agreed to pursue the matter further. At the Lab, I was told that I could not be offered a position unless I first resigned from my government job in New

Jersey. Although this process entailed some risk and meant leaving a job that had been enjoyable and satisfying to me, I decided to proceed.

As we prepared to leave, we decided that since we had another car that we had left back in New Jersey, we would leave the Corvair in California and take a commercial flight home. My group leader had kindly offered to allow us to put the Corvair, filled with many of our belongings, into his garage until we returned. When our return flight to the east coast landed in Philadelphia, our parents who picked us up couldn't understand what we had done with the car that we had driven to California. At this point, we had to break the news to them as to what we were planning to do.

Several days later, I returned to Picatinny and told my supervisor that I was planning to resign. I told him that I hoped that a permanent job at the Livermore Laboratory would be offered to me. With my formal resignation paper work in hand, I advised Livermore of what I had done. A few days later I received an offer letter from Livermore stating that I would be granted a position within the same division and group that I had been working in previously.

As I prepared to leave, the members of the group that I had been working with at Picatinny took me to lunch at a tavern. We all ordered a drink, and embarrassingly, I was the only member of the group who was asked to prove his age. At that point, I was twenty-nine years old.

We put the house in Succasunna up for sale and were fortunate to sell it very quickly to a person whose wife worked at Picatinny. We closed on the sale within a period of three weeks. In late May, we were ready to begin another driving journey in returning to Livermore, California. I should mention at this point that Dori was now pregnant with our third child and was due to deliver sometime in August.

The car that we were driving to California was a 1955 Ford Fairlane that I had maintained very carefully and was very fond of. It had premium tires on it, but the rears had less than half of their tread remaining. I judged that they would be adequate for the trip. As we proceeded westward, we encountered some heavy rain in Wyoming.

We were on a four-lane divided highway, cruising at about seventy miles an hour. We encountered a mild upgrade and a gentle sweeping turn to the left. As we entered the turn, I lost control of the car, and the car spun around several times on the highway before bouncing off a metal guard rail on the right side of the road. We hit the guard rail sufficiently hard that I expected that the car might break through and roll down a steep slope that was below the roadbed.

Although the car had front seat belts, my daughter Susan, who was riding next to me, was not wearing hers. I recall seeing her floating up towards the roof, as I attempted to hang on to her and bring her back down. Meanwhile, in the back seat, there were no seatbelts, and Dori's legs were bruised as she slid towards the front seats. I believe she still has a couple of scars on her shins from this encounter. A truck driver, who had witnessed the accident, stopped to offer assistance and to direct traffic around us. Fortunately, although my pride and joy Ford was damaged both in the front and rear ends where it had encountered the guard rail, it proved to be drivable.

We arrived in Livermore a couple of days later somewhat shaken, but fortunately uninjured. As we had previously, our initial stay in Livermore was at the Townhouse Motel. From there, we immediately began shopping for a permanent residence. As we looked at both new and used homes, we found none which satisfied our requirements. First, because we knew that we would soon have three children, we wanted four bedrooms. We also wanted a house to contain a living room, dining room, kitchen, and a separate family room. All of the available homes that we looked at failed to satisfy our requirements one way or another. We were also unhappy about the size of most of the building lots. They were small and all were separated from each other by solid fences to provide privacy.

One day, when we took one of the flying club airplanes for a local flight, I was amazed to see that out on the south side of town a person had a small airplane parked in his driveway. After we landed, we drove out to the area where we had seen the parked airplane. This area was on Marina Avenue where we saw two large ranch houses. One of them had the airplane parked on its property.

After contacting the owners of these properties, we found that the two owners owned in partnership a third building lot next door of one and a half acres. After some negotiations, we purchased that adjoining lot. Having made this purchase, our need for temporary longer-term housing caused us to return to the Biscayne Apartments.

We moved back in to the Biscayne and placed into storage most of the furniture that we had brought with us from New Jersey. We began the consideration of what kind of a house to build on our new lot, as well as building one we were able to afford. We attended a home show in the bay area and found a company that was offering pre-cut homes for what appeared to be a reasonable price.

After some deliberation, we decided to purchase one of the largest models that the company had available. It was a single-level ranch home, about ninety-nine feet in length, with the living space being about 2,250 sq. ft. (relatively large for that time period). It had four bedrooms, two baths, a living room, dining room, and kitchen, with a good-sized separate family room. The company allowed us to modify the plan to provide a two-car garage, with a larger foot print than what they ordinarily provided.

I think that we were a bit naïve as we entered the process of building this home. Basically, the pre-cut shell left us with having to provide a concrete foundation, plumbing, electrical work, and heating and air conditioning. As we added up the costs of the pre-cut shell and all of the other contracts, the bottom-line price was well above our initial estimates. Nevertheless, we proceeded with the project, with construction being started in the fall of 1961.

In the meantime, in August, Dori delivered our son Billy at St. Paul's hospital in Livermore. Although she went into the hospital with a bad cold, the delivery went quickly and mother and son were just fine. Our little two bedroom apartment in the Biscayne became a bit crowded. Dori and I occupied one bedroom. We attempted to put the girls in the other bedroom, but squabbling made it necessary to separate them. We ended up putting Linda in the bedroom and Suzie to bed in our bedroom. When we went to bed we carried Suzie out to the couch. There wasn't room for a crib, so Billy slept in his baby carriage in the kitchen.

As construction proceeded, we attempted to take on some of the minor work ourselves. We did all of the exterior and interior painting, as well as installing insulation. When the sheetrock was finally in place, Dori would take the children out to the house during the day and paint until she would go home in the late afternoon to prepare dinner. I would come home from the Lab, eat dinner, and return to the house in the evening to paint as long as I could with temporary drop lights.

In January of 1962, we were finally able to occupy our new house. Our heating and air conditioning contract was done through the local Sears store. We had opted for a two-unit, five-ton heat pump. The heat pump was selected because there was no piped gas available on Marina Avenue at that time; therefore, most of the existing houses there were heated with propane. The heat pump was designed so that the heat exchanger for the interior air was located in a concrete-lined pit under the house. The heat exchanger had been placed there prior to the flooring above being installed. The heat exchanger for the outside air was located on the back patio.

As we started through the winter rainy season, I noticed something strange occurring in our bathroom. I was getting lots of steam on the mirror while I shaved. Also, I noticed that the heater blower was operating more slowly and sounded different. I proceeded to investigate by removing the access hatch to the crawl space located in one of the closets. When I shined my flashlight toward the heat transfer unit, I was appalled to see that the pit contained approximately two feet of water, completely inundating the heat transfer unit.

We were without heat for several days as I broke out a circular section of the pit's concrete floor. I then proceeded to excavate about a two and a half foot well, in which, after lining it with concrete, I installed a submersible pump. Since the ground behind the house sloped downward, I was able to bring the discharge line from the pump to the back patio and let the water run away.

With this system in place, we had no further problems that year. However, in subsequent years, as heavy rains began, we would listen intently for sounds of an operating pump. Unfortunately, after the dry season, the pumps would often not restart automatically. I learned the hard way that it was necessary to always have a new pump at the ready.

Changing the pump always provided a sight to be seen situation. There was no point in going under the house with a full set of old clothing on because the ground under the house was very muddy. To carry out the operation, I would put on a bathing suit, crawl down through the mud, exchange the pump, and climb back out, covered from head to toe with wet mud. Dori would line the hallways with newspapers so that I could get to the shower without destroying our champagne-colored carpet.

In other ways, the house was somewhat ahead of its time. Not only with respect to its heat-pump heating system, but because all of the lighting outlets were controlled by low-voltage, push-button switches. Due to this, it was easy to wire the house so that, from a single switch, one could control a number of 110 V circuits.

CHAPTER 10

WHILE OUR NEW HOUSE GOT lots of attention, I was attempting to establish myself in B Division at the Laboratory. I continued to work within the same group that I had worked with while I was on assignment with Picatinny. Our focus was advanced concepts, which were particularly applicable to nuclear weapons that might be utilized in the defense of Western Europe.

As a part of this work, I took my first trip overseas, accompanied by a physicist from A Division and a senior engineer from W Division. The U. S. contingent also had representatives from Los Alamos and Sandia. Our interest was primarily focused toward the application of atomic demolition munitions, which might be employed on choke points to slow the advance of fast-moving Soviet armor.

We were taken to points along the highway in the Fulda Gap region and by helicopter were shown some of the mountain passes through the Alps in northern Italy. I still recall being at a meeting in army headquarters in Frankfurt where General Creighton Abrams was in attendance. Abrams, at that time, was a highly-respected four star general. Later, our heavy tanks were called the Abrams Tank in honor of the man.

In the Fulda area, we were given a classified briefing outside regarding ADM plans. Curious German civilians got out of their cars and stood along a nearby road, trying to observe what was going on. The army had set up a perimeter screen to keep the Germans out of ear-shot distance.

Although our efforts on advanced concepts were interesting, I wasn't really involved with the basic calculations and experimental activities that was the bread and butter part of B Division. The group leader I had been working for was reassigned to A Division. I was transferred to another larger group that was responsible for the calculations and experimental activities associated with a certain class of fission weapon designs.

Within this group, I was running many hydrodynamic, neutronic, and yield calculations. These calculations were utilized to determine optimums in terms of performance and efficiency. The person that headed this group was probably one of the best experimental hydrodynamicists in the business. After a year or so in this group, I was fortunate to be able to work with a senior designer who had transferred from Los Alamos. His expertise in theoretical one-dimensional hydro was outstanding. While working with him, I learned a lot.

For several years B Division had been dismayed over the poor performance of one configuration of one of our small nuclear devices. Using calculational techniques that I had learned and employing those techniques against some devices that had performed better, I came up with a new design that I was sure would rectify the problems of the poor performer.

After many calculations, an inert version was built and fired at our hydro-dynamic testing facility at Site 300. With all aspects of the design looking good, a nuclear test of the item was scheduled. For the nuclear event, as with all nuclear events, I prepared my pre-shot which gave predictions of the experimental values to be expected. One of the main predictions was, of course, the nuclear yield. My pre-shot value was 3.8 kilotons, plus or minus 10%. The shot was fired, and when the measured yield was finally reported, it was 3.8 kilotons, plus or minus 10%.

Because the design was in a physics domain that had given

difficulties in both the U. S. and British programs, the test success now provided a capability in this certain class of devices that had been so difficult to achieve. The design techniques that I had employed continued to be utilized for systems that we would develop in the future.

In 1967, there was a requirement for a specially-tailored fission primary needed to drive a large high-yield thermonuclear system slated for the Spartan Anti-ballistic System. My primary design for that application was highly successful. It was fired tens of times in various live Nevada nuclear tests. Also, in 1967, the group leader for my group became the division leader for B Division. At that point, I assumed the position of group leader of the group in which I had been working.

During the next five years, serving as group leader, I enjoyed what was perhaps the most satisfying and productive interval of my entire career. Our group cranked out many designs that were unique and provided potential military capabilities beyond those which had been available before.

CHAPTER 11

RETURNING TO MY PERSONAL LIFE, I embarked on another hobby by purchasing a used small cabin cruiser (17 ½ feet). I bought it without a motor or trailer. I went out and bought a new 75 h.p. Evinrude outboard, which I hooked up and mounted on the transom. A new single-axle trailer was also purchased. Now the Scanlin family was, in a small way, in the boating business. For about two summers, we enjoyed our little boat, exploring the Delta, and camping and swimming off of remote beaches. For vacation, we had two weeks available. We trailered the boat to Lake Shasta, where we camped onboard and enjoyed swimming and water skiing in the warm clear waters of that lake.

Later, we decided that we were interested in more challenging boating. We decided to take our little boat north to the Puget Sound area. In preparation for that trip, we were fortunate to have installed full canvas over the back section of our mini-cruiser since we did encounter some rain.

We planned to have our three children sleep in the V-berth in the forward cabin. Our son, who was still quite small, would sleep on the filler-cushion which covered the marine toilet beneath. Dori and I were to sleep in sleeping bags on air mattresses we would blow up and lay on the floor in the cockpit in the rear.

With everything so prepared, we headed toward Seattle, towing the boat behind a 1957 Ford station wagon. We stopped at campgrounds along the way and used the boat as a travel trailer. On the road, the children slept in the boat, while Dori and I slept on a mattress pad in the back of the station wagon.

When we arrived north, our launch point was Anacortes, Washington. A crane, with a sling, lifted the boat from the trailer and dropped it into the water. The car and trailer were left in a parking lot, while we spent about a week cruising the northwest waters. Fortunately, we had marine charts available for navigation, but we had almost no other equipment. The boat had no depth sounder, nor did it have a radio, but it did have a magnetic compass. Thus, we were very much on our own in unfamiliar waters. Being a pilot, however, was very helpful in that both Dori and I were familiar with navigating from printed charts.

Our trip went well, and since our boat was rather fast, we were able to cover a lot of area in short order. Along the way, we befriended an older couple who had boated in the area for years. They were able to point out to us points of interest that we should explore. In fact, we followed them into a couple of ports where they had suggested that we go.

On the day that we headed back to Anacortes, we ran into a dense fog bank several miles before reaching the harbor. Since visibility was probably less than thirty feet, we shut down and let the boat drift while we listened to fog horns. We knew that there was ferry boat traffic in the area, and became very concerned because it seemed that one was approaching in the fog. At any moment, we were afraid of the possibility that a large gray shape would loom into view and swamp our little cruiser. Fortunately, that did not occur, and by homing in on the fog horn sound, we were able to move slowly forward and break into the clear just outside the Anacortes harbor.

On our return trip, we stopped briefly in Seattle and went up the Space Needle for breakfast. The Space Needle was very new at that time because the Seattle World's Fair had recently ended. Our breakfast was expensive, which resulted in us being almost without cash for the remainder of the trip. Credit cards at that time were virtually non-existent. It was near the end of the month so our bank account at home was nearly depleted.

In order to obtain the cash we needed to get home, we went to a local bank and cashed a personal check. With only several days until another deposit would be made, we felt that writing the check was probably O.K. Nevertheless, had there been fast electronic means of checking account balances, our check would have bounced.

After we returned home, using our little boat in the Delta seemed very tame compared to the boating that we had done in the northwest. Furthermore, we realized that our children were growing and the boat that we had was soon going to be inadequate for camping trips such as the one we had just completed. Our little boat was put up for sale.

CHAPTER 12

WITH THE BOAT GONE, MY interest returned to flying. I became the maintenance chairman for the Flying Particles Club. In that position, I had to arrange for annual inspections for the airplanes we had and schedule routine maintenance, such as oil changes and fixing minor problems, as reported by our flying members.

As I recalled earlier, our next door neighbor had kept his Piper Tripacer in his driveway. To fly in and out from his home, he taxied across the street and used a strip located on another neighbor's property. That strip was narrow and went up and down, almost like a roller-coaster. Also, it was oriented such that the prevailing winds were always cross winds for take-off and landing.

I flew with him once out of that strip, and, believe me, the take-off and landing was a bit hairy. The fellow that I flew with had been a B24 pilot during WWII. I believe that he had developed an attitude of invincibility. I never flew with him again. I heard later on that he had crashed other airplanes in other locations.

In addition to the strip across the street, there was a good-sized hay field belonging to a farmer located beyond our back fence. It was reasonably level and oriented into the prevailing wind. As maintenance chairman, I decided to bring the Club Tripacer into the farmer's field. I opened up the back fence and brought the Tripacer up to our back patio for a cleanup and oil change.

The landing went smoothly enough, and, as planned, the oil was changed, the plane was washed, and some polishing was accomplished. Another member of the Club had joined me in this enterprise. When our work was completed, the other fellow took the left seat and we taxied down to the farmer's field for take-off.

The field seemed a little short and was slightly uphill. As we began our take-off roll, some of the hay was long enough that our propeller was mowing it. Our cleaned-up airplane was being covered by many stalks of the mowed hay. In addition to the uphill roll, the growth on the field slowed our rate of speed gain. As the airplane reached an air speed of about fifty miles per hour, we noticed that a barbed wire fence at the end of the field was only a couple hundred feet ahead of us. I called out to my buddy who was flying and said, "This thing has got to fly NOW." As he pulled back on the controls the plane staggered into the air, barely clearing the barbed wire fence.

We had been fortunate to have avoided a disaster. Hitting the barbed wire may have flipped the airplane on to its back, doing serious damage. Furthermore, on the following day, we received a call from the farmer who owned the field who did not appreciate what we had been doing in his field. My aircraft maintenance exploits at our house never occurred again.

Around this time a small group of us decided to leave the Flying Particles and purchase our own airplane in joint ownership. Six of us kicked in $1,500 each and paid $7,500 for a used 1960 Cessna 172. It was a clean airplane that we got some good use out of. Several of us, however, yearned for an airplane with more performance.

Accordingly, we found a 1951 Navion Super B, which had a 260 h.p. Lycoming engine. It was a complex airplane in that it had a controllable pitch propeller, retractable landing gear, movable cowl flaps and hydraulically activated flaps. The original Navion design was made by North American Aviation. The same company had produced the P51 Mustang during WWII. Thus, with its sliding canopy, there was a family resemblance to the Mustang.

Because of the complex nature of that airplane, it was necessary for us to engage an instructor to give us a very thorough check out before flying it on our own. Of the six owners, only three of us successfully completed the check-out process. I believe the other three owners were somewhat intimidated by the complexity of the Navion. It was the nicest airplane that I had ever flown.

At that time, the airstrip at the Nevada Test Site allowed private aircraft to come in and out. When I was involved in nuclear testing at work, I flew the Navion down there several times. It was a handy process since, at the completion of work each day, I would take the Navion from the Test Site strip down to the McCarran Airport in Las Vegas. Returning to the Test Site in the morning made for an easy trip.

I also flew the Navion into the Tonapah Test Range where we did testing of inert nuclear artillery shells out of 8" and 155 mm canons. Operating this complex airplane, with our small group having limited funds, was a greater expense than we should have undertaken. With misgivings of several of us, we finally sold the Navion to a dentist in Redding, California, and rejoined the Flying Particles Club.

Davy Crocket – 1960

Arthur Fleming Award Photo in 1957

Suzie, Billy, and Linda in 1962

Navion – 1964

Bill in 1977

Working in Office – 1977

First Boat – 1963

CHAPTER 13

THAT CLUB HAD NOW EXPANDED and had several late model Cessna 172 aircraft available. We had done lots of flying in and around northern California with our family, such as to the Nut Tree, Half Moon Bay, and the Napa Valley. We thought we would take our next vacation flying one of the club's aircraft back to Philadelphia to visit our parents.

In preparation, we obtained all the necessary charts and signed up to fly one of the newest of the three 172s. As our departure day approached, a problem developed with that aircraft and we were switched to a used one that the club had recently obtained.

In order to minimize the weight that we would carry, we sent, by way of Greyhound, a box containing clothes and other items that we would need for a two-week stay back east. We were thus able to pack everything we needed for a couple of over-night stays in some rather small vinyl bags that we were able to stuff under the seats. Since the plane only had four seats and there were five of us, a jump seat which fit into the baggage compartment, was acquired. Our son Billy, who was the smallest and lightest of our kids at that time, rode the seat in the baggage compartment.

As you travel eastward, thunder storms tend to develop in the afternoon. Therefore, our plan was start out early every morning and fly until about 2:00 p.m., at which time, we would find a motel and let the children swim in the motel pool to let them work off some steam after having been confined in the airplane so long.

At about 6:00 – 7:00 a.m. in late July, we departed from the old Livermore Airport and headed eastward. After a refueling stop in Elko, Nevada, our next stop was to be Ogden, Utah. As we landed there, we could see that the Wasatch Mountains just to the east of the city were obscured by thunder clouds. As a result, our first day's flight ended at that point. Early the next morning the mountains had cleared and we were able to proceed farther east.

As we got beyond the Rocky Mountains, the terrain flattened. While we were cruising at about 7,000 – 8,000 feet, we encountered some white puffy cloud buildups, with tops just slightly higher than our cruising altitude. The openings between the clouds began to narrow so I began to climb, attempting to get over the top of them. Unfortunately, the clouds were rising more rapidly than our airplane would climb.

At that point, I noticed that my air speed had dropped to sixty miles an hour, which was uncomfortably close to the airplane's stalling speed. It was necessary for me to let the nose drop to regain flying speed. As we did, the airplane entered the clouds and all visual reference was lost.

It was fortunate that, although I am not instrument rated, I had done some instrument practice, with an instructor, just prior to my departure. I was, therefore, confident that I could keep the airplane under control as I made the prescribed 180 degree turn to exit the cloud. After flying back several miles, I noticed that the broken clouds didn't extend to the ground. I was, therefore, able to spiral down to an altitude of 1,000 – 2,000 above the ground and proceed beneath the broken layer.

Our next stop was Iowa City, Iowa. From there it appeared that we would be able to make Philadelphia by the afternoon of the next day. We got an early start out of Iowa City and were able to make it to our next fueling at Mt. Vernon, Ohio. On that leg, I pushed my fuel supply a bit more than I should have. After we topped off, I found that I had only five gallons of fuel left. In other words, only about thirty minutes of time before empty.

Proceeding eastward, we ran into a ridge of storms on the west side of the Appalachian Mountains. It first appeared that we might not be able to go on. However, I probed the cloud ridge heading toward the northeast, and found what appeared to be a hole through the line to the other side. Unwisely, I decided to go through. There was turbulence, and apparently we had picked up an electrostatic charge as our VOR indicator flopped around wildly. Fortunately, after a few minutes, we popped out into clear weather on the other side. Being somewhat shaken, we decided to land in Altoona, eat some lunch, and unwind.

Our next stop was our final destination. In the late afternoon, we landed at historic Wings Field in the suburbs of Philadelphia. As we taxied into our parking area, we saw our parents sitting on a bench waiting for us. I think everyone was relieved that we had made the trip across the country in two and a half days, in spite of having encountered a couple of weather issues. That evening my dad took us to a very nice restaurant where we all wound down, enjoying a dinner that seemed to be one of the best I had ever had.

During our two weeks on the east coast, we explored some of the sites around Philadelphia, visited old friends, and drove to Townsends Inlet on the Jersey Shore to enjoy a day at the beach. The two weeks went by quickly and it was soon time for us to start our long flight back.

Our early morning departure out of Wings Field brought clear but very hazy conditions. As we climbed to cruising altitude, we noticed that, due to the haze, there was no horizon. The only ground that you could see was a small area directly below the airplane. Although it was legal for me to fly in these conditions, it was still necessary for me to fly the airplane by reference to the instruments only. As the morning wore on, the haze diminished.

Nearing Pittsburg, Pennsylvania, I decided to increase my cruising altitude. As I did so, the engine suddenly became very rough. I went through a series of checks to see what might be wrong, but it was neither carburetor ice, nor a problem with the fuel, since both tanks were still nearly full. I called Pittsburg Center on the radio and explained that I thought that I might be losing the engine. They cleared me for an emergency approach into Akron Canton Airport, which was directly ahead. As we landed, the airport fire equipment followed us down the runway.

I taxied the airplane to the airport fixed-base operator and explained to the mechanics what had occurred. They were very thorough, inspecting everything that could seemingly have caused the problem. The spark plugs were removed, cleaned, and tested. The ignition wires were tested. The magnetos were opened and inspected, the fuel was drained, and the fuel screen was cleaned to make sure no foreign material or water was present. Everything was reinstalled and a static run-up of the engine was performed. Everything seemed to operate perfectly. Whatever the problem was, the mechanics felt that they had taken care of it.

We took off again and continued our western journey. As I got over Illinois, I attempted to gain some altitude, and lo and behold the same problem occurred. Lowering the nose and ceasing the climb temporarily rectified the problem, but as a precaution we landed in Kankakee, Illinois, and proceeded straight to the Cessna dealer on the field. We explained to the mechanic there what had been done back in Ohio.

The mechanic did some more checking. He had observed a bulletin which spoke of a potential problem with the carburetor. He proceeded to disassemble the carburetor and install a modification kit as the service bulletin had prescribed. Having done this, a static run-up of the engine was again performed. Everything behaved perfectly. By this time, we felt quite uneasy about the airplane we were flying as we proceeded westward.

Sure enough, we had not gone very far when the rough engine problem again reared its ugly head. This time, we were near Omaha, Nebraska. We knew that this problem needed to be resolved before taking on the Rockies and the Sierra Nevada farther to the west. Landing in Omaha, I again took the plane to the Cessna dealer. In so many cases such like this the mechanics have some suspicion that it is the pilot, not the aircraft, who is causing the problem.

Once again, I explained to the head mechanic what had been done and what we were experiencing. He said that he would like to fly with me, and at this point I was only too happy to have him do that. He flew the left seat and, climbing away from the airport, the engine again became very rough. We returned to the airport, and we discussed what the next step should be.

Since all the external components, like the spark plugs, the magnetos, and the carburetor had all been worked on with no apparent change, this mechanic felt that it was necessary to tear into the engine itself. Because of the mountainous terrain that lay ahead, I felt strongly that this was the thing to do. By phone, I reported the situation to the president of the flying club. He agreed that anything necessary to rectify the problem ought to be done. I also told him that I wanted to leave the airplane in Omaha while the airplane was being repaired and return my family to California on a commercial flight. The president of the flying club gave me permission to do that.

The next morning before leaving I stopped by the shop and checked on the engine disassembly. One of the mechanics showed me that, when he passed the proper-sized reamer through the valve guides, it cut a small amount of metal. If the guides had been properly sized, this cutting of metal should not have occurred. Apparently, when the engine had been overhauled about two hundred hours before, a slightly smaller-sized reamer had been used. As a result, when we attempted to climb, the valve-stem temperature would rise slightly which caused expansion of the valve stem, preventing the valves from closing properly. One of the other club members returned to Omaha about a week later and brought the plane back to Livermore.

Back in Livermore, the plane never gave any trouble again. Although our long light plane trip was successful, it provided a bit more excitement and apprehension than my wife and family cared for. We decided that for such long trips commercial flights were the way to go. Our activity with the flying club diminished.

CHAPTER 14

OUR HOUSE ON MARINA AVENUE, although large and comfortable, had a continual string of annoying and expensive problems. I recounted earlier about the water problems associated with the under the house heat pump. We also had problems with our well, which was required to provide us with our domestic water supply. That well had a submersible pump suspended about two hundred feet down. There was some sand in the well, which shortened the life of the pumps.

Our sewer system consisted of a septic tank and a large drain field at the back of the property. Due to poor percolation rates on our property, the drain field was necessarily quite large. The drain field consisted of two hundred feet of trenches about two feet wide, filled at the bottom with about three feet of crushed rock. The trenches were fed the water from the septic tank with perforated pipe imbedded in the crushed rock. From time to time, the sewer line from the house would plug up. Some very nasty episodes of sewage backing into the bath tub which our children used had sometimes occurred.

With these problems, and because our children had to take long school bus rides to their schools in Livermore, we started to look at housing in Livermore itself. As we looked, we came across a house in Granada Woods, located on Escondido Circle. It had been one of three model homes Mesud Mehran had built to start off the Granada Woods development. It was built in 1963 and was occupied for only about a year by Mehran's building superintendent and his family. As we had always insisted upon before, the house had four bedrooms, a living room, dining room, kitchen, family room and two and a half baths.

Overall, it was slightly smaller than the house that we had on Marina Avenue, but it had lots of nice finishing touches that made it very attractive inside. Since it had been a model, it was beautifully landscaped and had a swimming pool, which our children thought was great. We sold the house on Marina Avenue and moved into the house on Escondido Circle in 1967. We happily remained there for the next ten years.

CHAPTER 15

AS OUR INTEREST IN FLYING diminished, our interest in boating returned. We began shopping for a larger trailerable boat, which could provide camping accommodations for a family of five, now consisting of three children who were approaching their teenage years. We shopped and looked at many different models in the size range of twenty-one to twenty-four feet in length. I happened to see a magazine article about the test of a 23 ft. boat made by the IMP Company. The test report came from a dealer down in southern California. We drove down there and looked at the IMP, along with Fiberform and Bayliner boats, of similar size and configuration. We settled on the IMP because of its high-quality construction and finish.

The model we purchased had two V berths in the forward cuddy cabin, two sleeper seats in the cockpit area, and a berth across the engine compartment at the stern. The boat rode on a dual axle, heavy-duty trailer. This boat performed very well as we used it around the Delta. In one trip we towed the boat to southern California to the Dana Point Marina. We used that as a jumping off point for making a run out to Catalina Island, twenty-six miles off the California coast. The boat had only a single engine, and, similar to the seventeen and one half foot cruiser we had owned before, it had minimal equipment. However, in addition to its magnetic compass, it did have a depth sounder, but no radio and no other navigational aids.

Because our trip to the northwest six years before had been so enjoyable, we elected to pursue that venue again. On this next trip, we had some good friends Lorraine and Don, who had a Bayliner, which was similar in size and configuration to ours. Our friends had three children, two girls and a boy like us. As we had done previously, we launched the boats at Skyline Marina in Anacortes, Washington. Since we had heard and read so much about it, our objective was to explore the Princess Louisa Inlet. Along the way, we made several stops at marinas where we boat camped off the boats and on the docks. In the evenings, we would off-load some of our paraphernalia and set up a propane stove on which we would cook our evening meal.

Getting to Princess Louisa is a fairly long run (about 50 miles) off the main channel. Although the channel is somewhat lengthy, it is impressive because the steep mountains descend into a fiord-like waterway. Numerous small waterfalls could be seen as the snowpack high above melted in the summer warmth. Before proceeding into the final stretch of the inlet, it was necessary to pass through some tricky rapids. Traversing the rapids at the wrong time had been known to put boats on the rocks. Since our small boats were fast and maneuverable, it was not necessary to wait for fully-slack water in order to negotiate the rapids.

As we approached the end of the fiord, we were treated to a spectacular view of Chatterbox Falls. Chatterbox Falls was much larger and more impressive than the falls we had previously seen. We could see the water feeding the falls, dropping down cliffs several thousand feet above. As we approached to the right of the falls, there was a long dock consisting of logs tied together. Some boats were already there, but there was space for us to tie up for the night. All of us enjoyed hiking up along the edge of the falls while our four blonde daughters enjoyed washing their hair in the sweet fresh mountain water.

After dinner, we went to bed tired after the fairly-long run we had made that day. During the night, we were awakened by our eldest daughter Suzie, complaining of sharp pains in her lower-right abdomen. Such pains were classic symptoms of appendicitis. We were all very concerned about what to do but knew that it was unsafe to attempt to exit the inlet in the dark.

In the morning, we encountered a medical doctor who was on another boat tied to the dock. He examined Suzie and told us that, if she were his child, he would head to the hospital as soon as possible. Reluctantly, we left this most beautiful spot and headed for the town of Sechelt where we were told that a significant hospital existed.

After making the long run down the fiord, we made a left turn into the Sechelt Inlet, first having to negotiate significant rapids. We had no time to wait for slack water, and even though the fast-moving water looked quite wild, we went full bore right through the rapids. Our friends in the other boat waited to see what happened to us before they proceeded. After we cleared the rapids, our excitement didn't end. A seaplane buzzed low over us and made its landing not far ahead.

Once having docked at Sechelt, we got Suzie to the hospital. As she lay on the examining table we were somewhat embarrassed by the fact that the bottoms of her feet were black from walking around in her bare feet. The doctors and nurses were very kind to us and felt that Suzie needed to stay in the hospital overnight to see if her condition worsened. Since we didn't have access to a telephone, we gave permission in advance for the doctor to operate to remove her appendix if it was deemed necessary.

We felt that we were leaving her in good hands as we returned to the dock to spend the remainder of the day and the night. As we returned to the hospital the next morning, we were gratified to find that Suzie's pains were gone and that the low-level fever that she had had disappeared. We were all envious that Suzie had had a good hot shower and was nice and clean. We were pleasantly surprised that there was no charge at the hospital for Suzie's stay, in spite of all the tests that had been done and the good care she had received.

On our return trip, one of our final stops was in Victoria, British

Columbia. Entering that busy harbor was exciting as we dodged heavy boat traffic and sea planes landing and taking off. We were fortunate to find dock space immediately in front of the beautiful Empress Hotel. We were impressed by the beauty of the overall water front area, which had lovely baskets of blooming flowers hanging from all of the lamp posts. We enjoyed visiting the wonderful nearby museum and the impressive parliament building on the south side of the harbor. The remainder of our journey was uneventful, in that we had no problems towing the boat home with our 1965 Mercury station wagon.

CHAPTER 16

ANOTHER GREAT JOURNEY WE ACCOMPLISHED while we had the 23' IMP Inca was a trip to Lake Powell. Since the trip took place during August, we encountered very hot weather in the southwest. Our Mercury tow vehicle tended to run hot pulling more than six thousand pounds of boat and trailer. When we got to Lake Havasu in the afternoon, we decided to spend the afternoon swimming in the lake and proceed onward only after the sun had gone down. Late in the evening we found a place off the main road to park the rig. Using the boat as a camper, we climbed up into it and went to sleep.

When we got to Lake Powell, we used Wahweap as our launch point. Although it continued to be very hot, we found Lake Powell to be very beautiful, providing us with boat camping experiences that were positively outstanding.

In one instance, we encountered a canyon that was so narrow that we were afraid that we might not be able to turn the boat around. Thus, if it became any narrower, we would have to back out. The sheer rock walls extended almost vertically upward, roughly three hundred feet above us. As we proceeded forward, we were rewarded by entry into a fantastic overhanging rock grotto. Within the grotto, there was even a small sand beach where we were able to secure the boat and spend the night. We were completely alone in there. It was so quiet that it seemed that we should speak to each other only in whispers.

At the time of this trip, Lake Powell was still in the process of filling as the Colorado River fed into it. As we proceeded farther up the lake, we found that we were able to explore a natural rock bridge, known as Rainbow Bridge. Because the water level was still down, we had to hike up to the bridge in temperatures well over one hundred degrees. When we got to the base of the bridge, we encountered a friendly young female national park ranger. She provided us with a cold drink and gave us useful information about points of interest in the lake. Later, we invited her back to the boat for lunch.

Our entire visit to that lake was truly magical. We never returned and had always felt that we would never have been able to replicate the wonderful time that we had enjoyed there.

CHAPTER 17

IN THE TIME FRAME OF 1971 and 1972, some of my colleagues at work were beginning to run. I decided that for the potential physical fitness benefits of running I would start to run too. It seems almost unbelievable that at that time it was necessary to go to a special sporting goods store to obtain a decent pair of running shoes.

For some time my running stayed with fairly short runs of about two miles. Later, I increased my distance to four-mile runs, with an occasional run of eight miles or so. I became so dedicated to running that when I traveled (most frequently to Washington, D. C.), I ran irrespective of weather conditions. I can recall running in the dark in temperatures below freezing and with ice and snow on the ground. I also ran overseas when traveling for work or when vacationing. Memorable runs took place in Paris, Athens, Egypt, and even Bangkok. I continued this dedicated running for a period of over thirty years.

CHAPTER 18

DURING THE EARLY 70S, MY work as head of a nuclear-design group continued to be very rewarding and satisfying. A special fission primary design that we were developing for the Spartan missile warhead had been tested satisfactorily a number of times. To complete the Spartan warhead development, it was decided that a full-scale, full-yield experiment with that warhead should be accomplished.

The planned yield for that experiment was high enough that it could not be conducted at the Nevada Test Site because the seismic effects on the City of Las Vegas would have been excessive. Another test area near Tonopah, Nevada, was explored as a potential test site. Due to the high yield, the required containment depth was in the order of five thousand feet. The temperature at that depth was very high, resulting on thermal effects on that warhead that would have been undesirable.

Test borings at one of the islands in the Aleutian Chain, namely Amchitka, showed that temperature at the desired depth would be acceptable. Therefore, a temporary high-yield test site was established there. The high-yield experiment planned at Amchitka was named Cannikin. Cannikin's emplacement hole was approximately eight feet in diameter and over six thousand feet deep. At the bottom of the hole, a room was mined so that when the device was emplaced, the early phases of the explosion would not be influenced by effects created by the adjacent walls.

I traveled to Amchitka about a week ahead of the planned firing date. I was to observe some of the final test preparations. I was also there to monitor the pretest functions of the fission primary, which was my responsibility. In that week while I was there, all kinds of activity was taking place. There were visiting congressmen to be briefed. There were law suits being filed in Washington to stop the experiment. The chairman of the Atomic Energy Commission arrived with his wife and two young daughters to demonstrate to the outside world that he considered the execution of the experiment to be safe.

There was one episode that I recall vividly that must be considered among the best practical jokes ever perpetrated. Due to the extreme depth of the emplacement hole, ground water was continually seeping in to the cavity at the bottom. In order to remove this water, high-power electric pumps were used to transfer the water to the surface. After the nuclear device had been placed in the hole, and after the hole had been stemmed, the success of the experiment hinged on the ability of the pumps to remove the incoming ground water.

As part of the emplacement, high powered quartz lamps were located in the cavity. These lamps, with a remote TV camera, enabled the test staff to have a visual indication of conditions in the test cavity. Because the quartz lamps had a limited life, they were typically turned on only once a day to perform visual monitoring. Due to the critical nature of the threat of pump failure and the flooding of the experimental cavity, it had become customary for several days prior to the shot for me, along with others, to check the conditions in the cavity.

Base camp, where we were housed, was roughly twenty miles from the shot point. The control point was about an equal distance from the shot point at the opposite end of the island. At the shot point, there was a small trailer known as the red shack where electrical terminations to the device down the hole were located. It was also where the down-hole TV monitor was mounted. As the shot date grew near, a helicopter was at our disposal to monitor activities and to brief visiting dignitaries at the control point.

On the way back to base camp we would stop at the red shack. At that time the quartz lights down in the shot cavity were turned on. Each time we did that we saw the white weapon casing hanging there in the center of the shot chamber. On one trip back to the base camp, a radio call came into the helicopter requesting that I come immediately to the red shack because there was something that I should see. Due to the tone of the radio call and the seeming urgency of it, I was somewhat concerned.

The helicopter landed adjacent to the red shack. As I entered the red shack, the engineers in there said, "Look at this." The TV monitor went on and there was the image of the white warhead casing as it had typically been. Suddenly, water began to rise in the bottom of the cavity, quickly engulfing the entire warhead. Since precautions to make the casing entirely water tight had not been made, the rising water would have meant disaster for the entire experiment.

Suddenly, a giant hand reached into the cavity and pulled the warhead out. At this point, I can only imagine what my expression must have looked like. After much laughter by everyone in the room, I realized that it was all a hoax and that the engineers had used a small model of the warhead suspended in a goldfish bowl. Of course, as it finally turned out, there never was a failure of the down-hole pumps.

On the day before the shot, an extremely violent storm hit the island. We experienced virtually horizontal heavy rain along with wind gusts that sometimes well-exceeded one hundred miles an hour. At base camp, I had gone outside to walk to the dining area. I was barely able to stand up because of the extreme winds. Fortunately, on the day of the shot, the storm went away and final preparations for the experiment were completed.

On the planned shot day, litigation against the test had failed, and authorization to proceed with the experiment was received. That morning I was transported to the control point where monitoring instrumentation was located. I was stationed behind an oscilloscope which read out an important pre-firing bit of data. If for some reason this data was not correct, the count down to zero time would have been halted. Because that data was so critical, the chairman of the Atomic Energy Commission, James Schlesinger, stood behind me, looking over my shoulder.

As the count-down progressed, the data on my monitor looked fine. As zero-time occurred, the instrumentation trailer we were in rocked and rolled more severely than any earthquake I had ever experienced. The good news about this shake was that the experiment was apparently a success. At that point, we didn't know exactly what the experimental yield was, but due to the seismic event that we had felt, we knew that the yield must have been large.

That evening there was a cocktail party to celebrate the apparent success of the experiment. Later we found that various measurements indicated that the device had produced a yield of less than five megatons. Lots of other instrumentation had measured various forms of radiation emitted by the warhead and were also very successful. The high-yield test site at Amchitka was never used again.

CHAPTER 19

MY WORK ON THE SPARTAN missile system prompted another trip to a distant location. On this trip, I traveled to Kwajalein in the central Pacific Ocean to view other components of the Anti Ballistic Missile System (ABMS). At Kwajalein were located the tracking radars that were used to track incoming reentry vehicles from ballistic missile launches. To test the system, ballistic missiles with their RVs were launched from Vandenberg Air Force Base in California and targeted to splash down in the large lagoon formed by the atoll around Kwajalein. One of the major components of the tracking system was the giant phased-array radar installed on the nearby island of Meck.

My trip to Kwajalein began with a flight from San Francisco to Honolulu. From there, I boarded a Continental/Air Micronesia flight to Kwajalein, with an intermediate stop at the island of Midway. When I arrived at Kwajalein, I found that my luggage had not been on the plane. I had arrived on a Saturday and was told that the next flight was not due to arrive until the following Wednesday. I had arrived with only the clothes on my back and a few toiletry items in my briefcase.

Fortunately, there was a well-stocked PX on the island. The PX had been nick-named Macys. In the PX I bought enough shirts and underwear to last until Wednesday. I looked forward to my bag arriving by then. On Wednesday, I went to the airport to retrieve my bag. I was very dismayed to find that the bag was not on that airplane either. The next plane was due on Saturday. That plane was going to turn around and take me back to Honolulu because my tour was due to end on that day.

Back to Macys I went again, buying more clean shirts and underwear. On Saturday, once again the bag did not appear. By that time it seemed my only hope of finding the bag would be in Honolulu.

Interesting things were observed during my stay in Kwajalein. We boarded a twin-engine Caribou aircraft, which flew to Meck so that we could tour the phased-array radar. Meck is a rather small island and has only a very short runway for landing and departing aircraft. As we approached Meck, my view out of the plane window showed a runway that looked like a postage stamp. I wondered how the pilot was ever going to be able to stop within such a short length. Apparently, the Caribou is made for that type of service. It made both the landing and the later take off easily.

When the ABM facilities were built at Kwajalein, some of the natives who had lived there needed to be relocated to some nearby islands. We took a small boat out to visit one of these islands. When the natives were relocated, it seemed that the military had, in a move of seeming generosity, agreed to provide food and beverages for these people. As we explored on foot, it was sad to see outside each set of dwellings large piles of empty cans of all kinds, but mostly those of soft drinks. The people had become lazy and were obviously not having a healthy diet.

On one evening, we were told that a Minuteman ICBM was to be launched toward Kwajalein. If it was on target, its reentry vehicle was due to splash down in the lagoon. Witnessing the reentry of an RV is different and more dramatic than I had imagined. As the reentry proceeded, the entire sky lit up more brightly than any thunderstorm

I had ever seen. There was also a loud report as the RV broke the sound barrier when it encountered denser air.

While in Honolulu, I made further checks on what had happened to my bag. Apparently, on my trip out to Kwajalein I assumed that my bag had been checked through. It was for some reason off-loaded in Honolulu, going around and around on the carrousel with no one picking it up. It was then sent to United's unclaimed baggage center in Chicago. Once United understood my situation, they delivered the bag to my hotel in Honolulu within twelve or so hours.

CHAPTER 20

IN 1972, I LEFT MY group-leader position in B Division to become a member of the staff of the Associate Director for Military Applications. My assigned area of responsibility was in the area of tactical nuclear systems. The major thrust of tactical systems at that time was for the defense of Western Europe. The Soviets and their allies in Eastern Europe had made enormous investments in manpower and tanks that the west was unprepared to try to match. Western plans to equalize the mismatch included a potential to employ tactical nuclear weapons to thwart any invasion attempts.

Because employment of nuclear weapons on Western European soil had the potential to cause serious collateral damage in the area and to friendly civilians, new nuclear weapon designs were explored to minimize this damage. Ways to accomplish this included keeping yields in a lower range, improving delivery accuracy, and reducing the fission-yield component. The fission-yield component of the weapon was a major contributor to the radio-active fallout which occurs after an explosion.

New designs, called clean weapons, had as low a fission yield in them as possible. In addition, other materials in the bomb were selected so as to reduce induced radio-active products in the bomb debris. Other attributes we sought for tactical nuclear weapons were weapons having a dual-use capability. Dual-use means that, should nuclear release not be received, the weapons system could also have a conventional high-explosive warhead capability. Unfortunately, advanced missile systems, containing guidance to achieve good accuracy, means that they are more expensive. Thus, a sustained employment using missile borne high-explosive warheads would quickly deplete an expensive limited resource.

Tube artillery on the other hand is an ideal dual-use candidate. The shells are quite accurate, without a guidance system. Conventional shells are quite inexpensive and are typically produced in quantities of tens of thousands. As a result, they can be fired all day without significantly depleting their stockpile. Theoretically, they can transition virtually instantly if a nuclear projectile is made available at the launch site.

While I worked in the Military Applications Office, I spent a fair amount of my time working with the Army, trying to understand what the military characteristics of nuclear artillery shells ought to be to satisfy the Army's needs.

At the time, new nuclear warheads were being explored for use in both 8" and 155 mm cannons. For these systems, the major challenge appeared to be how to keep the yield low, to minimize collateral damage, while maintaining an adequate level of military effectiveness.

Another facet of tactical nuclear weapon employment involved the examination of response time. A request for a military mission could easily involve many minutes, or even hours, from the time of the request until the time when permission was given. The problem was examined by observing the movement of armor and other resources in large-scale exercises carried out at Fort Hunter Ligett.

As these movements were observed, the results were computerized. Once being computerized, response time and weapon characteristics could be evaluated as significant targets move across the battlefield. As these studies were carried out, once again tube artillery seemed to provide more of the desirable characteristics needed if ever tactical nuclear warfare ensued.

One nuclear design that appeared to be particularly appropriate was one having an enhanced neutron output. When the idea for such a warhead appeared in the press it was called the Neutron Bomb. The enhanced neutron-type system was particularly appropriate to a battlefield containing many enemy tanks. The reason is that tanks are quite hard to blast and thermal effects, but the crews inside the tanks can be vulnerable to radiation if the radiation is strong enough. The desirable attribute of the enhanced radiation warhead is that its prompt radiation effects are like a higher-yield weapon, but the blast and thermal effects are very much lower. Therefore, when used, tank crews inside tanks can receive a lethal dose of radiation, while fall out blast and thermal effects to nearby terrain and dwellings is much reduced.

After a couple of years of being involved with these studies of advanced warheads, I returned to B Division to take a position as program leader of a Livermore program to develop a modern 8" artillery shell, having an enhanced radiation capability. I remained as program leader of the eight inch development as well as being a group leader designing other tactical systems having special military characteristics. I stayed in those positions for several years.

CHAPTER 21

ON MY PERSONAL SIDE, MY family was enjoying the recreation possibilities that a twenty-three foot trailerable boat provided. However, we became interested in trailerable boats that could provide a greater degree of creature comforts. In early 1972, we had seen magazine advertisements for a twenty-eight foot cruiser named Land 'n Sea. These boats offered comfortable sleeping for five or six and had a complete bathroom with a separate stall shower. The galley had a two-burner propane stove and oven. The refrigerator operated three ways. It could run on 110 V, 12 V of battery power, or on propane.

The boat was only eight feet wide so it was legal to travel in all states without special arrangements such as wide-load signs. The boat was very much an amphibian in that it could be used as a self-contained travel trailer on land as well as being a self-contained cruiser when on the water. For land use, it had a fold- down stairway at the stern to allow easy boarding from the ground.

After visiting the factory, located in San Jose, we decided that this was the boat that we had to have. In due course, we sold the twenty-three foot Inca, and in December of 1972 took delivery of a brand-new twenty-eight foot Land 'n Sea. The particular model we ordered had a flybridge, containing a full set of controls, replicating those which were located below. For our boat, we specified two six-cylinder Volvo engines with outdrives. With this set up, the boat could easily cruise at over twenty-five miles an hour. Although it was quite large, we were still able to water ski behind it. The main disadvantage of the rig, however, was that the boat, with its trailer, weighed twelve thousand pounds, about twice what the Inca had weighed.

Our poor 1965 Mercury station wagon suffered mightily pulling this heavy load. As we used that Land 'n Sea on a number of long-distance trips, a number of weaknesses in our tow vehicle surfaced. The automatic transmission tended to run hot, causing it to fail several times. Rather than having expensive rebuilds performed, I chose to locate low-mileage used transmissions from wrecking yards, which I became adept at removing and replacing myself. Finally, in desperation, I decided to have one rebuilt by an outfit in Los Angeles, which specialized in building heavy-duty racing and RV transmissions. That final fix was successful.

We also found that the rear-axle differential would overheat, boiling out the lubricant, which, when it got on the exhaust system, created lots of smoke. The rear axles received lots of rebuilding attention also. As a result of the heavy tongue load that the boat's trailer imposed on the Mercury, we even broke wheel rims, resulting in air leaks. Fortunately, switching to some heavy duty highway-patrol rims, licked that problem as well.

One of the supposed features of Land 'n Sea ownership was that you could sell them and receive a commission from the factory. We decided to give this possibility a try by taking our nearly new boat to Las Vegas and displaying it for several days at an RV show in the Convention Center down there. The boat attracted lots of interest. We came very close to selling one, but never did.

We owned the Land 'n Sea from 1972 until 1980. During that interval we towed it to the Pacific Northwest several times. On one of those trips, we returned again to Princess Louisa Inlet. We also took that boat to San Diego where we attempted to go deep sea fishing with it to catch tuna in the Pacific off the coast of Tijuana. We used it in some of the northern California lakes. We had an especially enjoyable week on it on Lake Oroville. Lorraine and Don with their twenty-four foot Bayliner and their three children accompanied us on that trip as well as some others.

CHAPTER 22

WHILE WE OWNED THE LAND n' Sea, we became friendly with Astrid and Ray, a couple who had a Land 'n Sea equipped almost exactly the same as ours. We boated together in the delta and in Land 'n Sea outings in the bay sponsored by the manufacturer. They also trailered their boat to the Pacific Northwest and joined us for about two weeks while we cruised the waters north of Seattle.

In 1978, Ray retired in his early fifties from his job in the Stanford University Computer Center. Since both Astrid and Ray were of Scandinavian descent, Ray's retirement allowed them to consider implementing their dream of cruising around Europe. In the fall of 1978 they attended a boat show in London, where they purchased a 39' Camper-Nicholson Ketch-rigged sailboat. They sailed it to Oslo, Norway, where the boat spent the winter watched over by Astrid's brother, a Norwegian.

The next spring, they began cruising by entering the Gota Canal on the west side of Sweden. We were invited to join them for a couple of weeks by meeting them at Vadstena, about the mid-point of the canal. The objective was to follow the canal to the east side of Sweden, which would take us to the port city of Stockholm.

In June of 1979, Dori and I flew from San Francisco to Frankfurt, where we rented a car. From Frankfurt, we drove north, through Germany to Denmark. From Denmark, we crossed to Sweden on a ferry boat. In Sweden we proceeded northward to Vadstena. The driving trip seemed longer and more strenuous than we had bargained for. I don't understand why we didn't select a flight destination closer to where we needed to be in Sweden. My choice probably had to do with my familiarity with the airport, flight availability and car rentals in Frankfurt, since I had traveled in and out of there a number of times in connection with my work.

We arrived a couple of days early in Vadstena. We took a hotel room in a historic hotel that had been converted from a nunnery. There was an ancient castle in Vadstena. The moat of the castle had been developed into a marina. The overall setting was very picturesque. We were well-away from the busy tourist area so few people spoke English. We learned that we had better luck trying to communicate with young people since most of them studied English in school. Teenagers would giggle in embarrassment when trying to answer our questions in English.

In a couple of days, Astrid and Ray arrived in their sailboat. We stowed our things on board in the forward cabin, and began our cruise through central Sweden.

The canal went through a very pretty rural countryside. The leisurely pace of the cruise meant that I could sometimes run along the bordering path and keep up with the boat as it proceeded from lock to lock. A number of the locks had no operator so we had to open and close the locks ourselves.

The boat was named "Our Dream," with San Francisco as its home port. A great deal of interest was generated because people assumed that the boat had been sailed from its home port many thousands of miles away.

Astrid had fallen and injured her ribs early in their cruise. Her painful ribs necessarily limited her physical activity. Typically, Ray and I would handle the locking chores, Astrid would steer the boat, and Dori would hold the boat off from the side of the canal so that the boat wouldn't be damaged as the water rose and fell in each lock. Astrid at that time was a tall slender redhead. She made an interesting spectacle as she piloted the big sailboat through the locks.

One interesting occurrence was when we looked down and saw a highway beneath us. The canal was part of a bridge going over the highway. After traversing many locks and the eastern half of the canal, we arrived in Stockholm. Stockholm is an interesting old city with lots of important buildings. One is the town city hall which contains the large room where Nobel prizes are awarded. There are rivers and canals running through the city, which break it into a number of islands.

Being so far north and close to the summer solstice, it seemed that night never came. Even at 2:00 a.m., pulling aside the cover to the hatch in our stateroom, revealed a bright sky. Before we left the boat, we took it out into the open ocean and hoisted all the sails. With a moderate breeze , we enjoyed a great afternoon of sailing in the Baltic Sea.

Returning to Stockholm, we went to the bus terminal to find a bus that would take us back to Vadstena to pick up our rented car. With some difficulty understanding the Swedish signs, we finally boarded the right bus that took us back to the car. Retracing our route in Sweden, we again boarded a ferry boat that took us from Helsingborg in Sweden to Helsingor in Denmark. In Denmark, we spent a few hours exploring Copenhagen. At one point we parked the car in a waterfront area and explored on foot.

We returned to the car, and as we attempted to leave the parking area, our car was surrounded by about six young men. It looked as if they were up to no good. Although some stood in front of the car, I decided to move forward. I felt that I needed to be prepared to hit them if they didn't get out of the way. I think they perceived my determination and decided that they had better move. This episode left us with bad feelings about Copenhagen and Denmark in general. In all of our many travels, we have never encountered a situation that seemed so potentially threatening.

CHAPTER 23

A COUPLE OF YEARS LATER Astrid and Ray had their sailboat demasted so that they could take it south through the canals of France, with the objective of cruising in the Mediterranean for several years. They were successful in this venture and enjoyed the Med cruising while leaving the boat winter in marinas in various harbors.

Once again we were invited to join them while they cruised along the Yugoslavian coast. The plan was that we would join them at the port city of Split, a good distance north of the well-known city of Dubrovnik.

Our flights to and from Europe were through the airport at Rome. In Rome, we rented a car and drove north to Padua, through Trieste and south into Yugoslavia. On the way, we were scheduled to pick up a niece of Ray and Astrid in Zagreb. We took a hotel room and waited several days for the niece to arrive. While in Zagreb, it was hot. My early morning runs were made with me being shirtless, wearing only running shoes and shorts. As I ran through the city streets and central town square I must have been an unusual sight because people often turned their heads to gawk. We never saw other runners in Zagreb and the people must have been unaccustomed to scantily dressed men.

We finally gave up waiting for the niece and proceeded to drive through the mountains to Split where we were scheduled to meet Astrid and Ray. In Split we waited for Astrid and Ray to arrive. Each day we would check in with the harbor- master at the marina, but we got no indication of their arrival. After three days, we were almost ready to give up waiting and go off on our own.

What had happened was that Astrid and Ray had docked their boat in Trogir, a small town to the north, because rat infestation in Split was reported to be a problem. Each day Ray would check with the harbormaster in Split and the harbormaster would tell him that he had not heard from us. One day, when Ray checked in with the Split harbormaster, he noticed our name written on the corner of a note pad on the harbor master's desk. We had left the phone number of the hotel where we were staying. Ray drove to the hotel and we were soon on board their boat.

A day or so later the niece appeared, so with us all together, we finally got underway. We planned to make our way northward, running slowly and stopping frequently to enjoy the scenery. At one stop Dori and I hiked up a river and were treated to the view of a series of picturesque waterfalls cascading down a mountain.

At another stop, we tied up next to another sailboat occupied by some Europeans. To our amazement about half of those aboard paraded around in the nude, while the other half were in bathing suits. To us it would seem more natural if the crew had been all nude or all dressed. Our presence in the next boat didn't seem to bother any of them a bit.

Tying up in another harbor, we were very close to a row of houses along the waterfront. There must have been no plumbing in the houses because each evening, when the husband came home from work, he would wash himself using a faucet on a pipe coming out of the ground at the front of the house. As it became dark, the inside of the house was lit by a single bare light bulb hanging from a cord in the ceiling. It was clear that communism had not been kind to the working class in that area. Tito had recently died, but the communists were still in control. Most of the Yugoslavian people we encountered were sullen and unfriendly.

At another stop, we shopped at a local meat market. We wanted to have some hamburger that night. In the store we could see no ground meat that was of a quality that we could accept. We, therefore, asked the butcher to cut up some lean chunks of beef and put it through the meat grinder. At first, the butcher refused, indicating that it was unacceptable to grind meat of that quality. Finally, some of the other women in the shop persuaded the butcher to go ahead. He reluctantly ground the meat, slapped the package on the counter, and we finally got the ground meat that we wanted.

Our most northern destination was Zadar. At Zadar we left Astrid and Ray, taking a bus back to Split to retrieve our car and drive back to Rome. When we left our car in Split, we parked across the street from the police station. We explained to the police that we would be gone for a while and they agreed to keep an eye on the car. When we returned to the car, we immediately saw that some items had been removed while we were gone. First, the windshield wipers were gone. Also, the wide decorative plastic rub strips on the sides of the Renault had been pried off and were gone.

The time we needed to return to Rome was tight so because of that and our concern that the police may have played some role in the theft, we left without reporting the theft to them. When we got to the rental car agency in Rome, we explained that the car had had some items removed due to theft.

The rental car agent would not relieve us of our personal responsibility unless he had a police report. It was suggested that we go outside and report the theft now, even though the theft had occurred several days before in a different country. Outside we quickly found a policeman and showed him what had been stolen. He wrote up a report, which we returned to the rental car counter. That being done, everyone was happy, and we were relieved of any liability.

Astrid and Ray continued to use their boat in the Mediterranean for several more years. After that, they decided that they had done what they wanted to do in Europe and decided to bring the boat back to the U. S. The boat made it back and was finally sold in Annapolis, Maryland. A year or so later, Ray passed away from kidney cancer. Thankfully Ray had retired early and had enjoyed some very fine cruising before his death. He had after all been able to fulfill his dream.

CHAPTER 24

BACK AT THE LABORATORY, THE development of the new enhanced-radiation artillery shell designated W79 was successfully going forward. Several full-scale underground nuclear tests of the physics configuration worked as planned. Gun firing tests of complete projectiles showed a few areas that required redesign, but the fixes were successful and the project proceeded.

When the program's enhanced radiation capability was revealed, the press printed a number of adverse stories about the radiation capability. The negative connotation they developed was that it killed people while leaving buildings standing. Around this time, Jimmy Carter had become president. It seemed that, as a result of the adverse publicity, Carter decided that the enhanced radiation capability of the new projectile should be omitted. Provision for re-installment of the capability at a future date was to remain.

In about 1977, there was a change in B Division's leadership. A new division leader was appointed, and he selected me to be his deputy. At this point, I was carrying the work load of being deputy B Division leader as well as continuing as program leader for the W79 artillery shell. In addition at this time, I had taken on the responsibility for program leadership on a new 155 mm nuclear projectile called the W82. I was clearly a very busy person at this time, but I enjoyed every bit of it.

Although I was very shy as a youngster and hated public speaking as I became older, I had overcome this. I was called upon to give technical briefings to senior people all of the time. I got so that this part of my work became one of its more satisfying components. Supporting my new confidence in giving these talks, an employee evaluation I received around this time read as follows: "Another asset to the Laboratory is Bill's exceptional ability to give presentations to visitors. Bill is one of the nuclear explosives program's most able and reliable briefers, especially, when the presentation of a new idea is involved. Bill has been praised numerous times by DOD officers for his talks."

CHAPTER 25

THE FIRST GASOLINE FUEL CRISIS occurred in 1973. This meant that towing our large Land ‘n Sea boat became a problem, since, when towing, the fuel mileage of our Mercury station wagon tow vehicle was very poor. After looking around, we located a covered berth in Bethel Island. This meant that we could keep the boat in the water rather than having to tow it from place to place. When fuel was especially scarce, we would weekend on the boat without even moving it out of the slip. The Land ’n Sea remained at that dock for several years.

Back in the early 1970s, we had looked at some water-front property in a development called Discovery Bay. When looking at Discovery Bay at that time, I was not sure that continued development of those properties was going to be sustainable. Late in 1976, we went back for another look.

This time it appeared to us that Discovery Bay was on a threshold of a period where development might take off. After much soul searching, we decided to purchase a building lot and engage a contractor for the construction of a new home. Having a home in Discovery Bay would allow us to moor our boat behind the house on a waterway which provided access to the entire delta and even out to the Golden Gate and beyond.

The contractor we engaged to build our house worked with us closely so that we could get all of the features that we wanted in the house, as well as keeping the overall cost of the project reasonably under control. A key feature of the house was a large garage. It could accept three cars, with lots of room to spare for a good-sized work area including some power tools.

We sold our Granada Woods house in the spring of 1977. Temporarily, we rented a new unoccupied house in Discovery Bay. This was a good situation because we were able to closely watch the construction of our new home nearby. It also provided us with dock space for our Land 'n Sea, which could be docked behind the rental house.

There wasn't much work on the house that we could do ourselves because it would have interfered with the work the subcontractors were doing. We did, however, take on the job of supplying and installing floor tile in a 700 sq. ft. area of the first floor. With some help from our son Bill, Dori and I were able to complete the tile installation in two weekends.

Just before Thanksgiving in 1977, we moved into the new home. Thirty-three years later, we are still there. Because of its age, we have made a number of upgrades to it over the past ten years. Since we designed it ourselves, we are still quite happy with it.

As the boats we have owned have become larger and larger, we have had to increase the size of our dock. With respect to the boats, we traded in our Land 'n Sea in 1980 on a new 34' Fibreform Executive cruiser. This boat served us well during the early 1980s as we participated in a number of Discovery Bay Yacht Club cruises. Dori was Fleet Captain in 1983. Except for one cruise, we led all of the yacht club cruises that year.

We wanted to see how well the Fibreform handled open-ocean cruising. We went out the Golden Gate and took the boat to Monterey. On our return trip, as we proceeded north of Santa Cruz, we encountered some very rough seas with high wind. It was already late in the afternoon. We decided to anchor in a cove near Ano Nuevo.

We set the anchor and watched the nylon anchor line vibrate under the strain of holding the boat against about a thirty-mile-an-hour wind. As night came, we watched carefully lights located on shore in order to make sure that the anchor was not slipping. Also, we used the depth sounder with its alarm set to trigger a warning if the water became shallower than ten feet. Had the alarm gone off, it would have indicated that we were coming into the surf and could soon be on the rocks. The next morning we hit calmer weather allowing us to proceed farther northward up to the Golden Gate. We kept that boat until 1984 when we sought a larger boat, with more comfortable accommodations.

CHAPTER 26

OCCASIONALLY, WE HAD DONE SOME boating with some friends, Helga and Mark, who we had known since my earliest days at the Lab. Their very first boat was a 34" Taiwan-built trawler. They often used the trawler in the delta, keeping the boat tied to our dock. However, they usually docked their boat in San Rafael. After gaining experience, they wanted a larger boat and had ideas of using it in the Pacific Northwest.

About this time, a new line of boats had just started to be imported from Taiwan. These well-built boats seemed to offer an excellent value. The dealer selling them was located in Paulsbo, Washington. Helga and Mark sold their 34' trawler and purchased a new 48' Ponderosa aft-cabin cruiser. When the boat arrived in Paulsbo, Dori and I were invited to join them for the boat's maiden voyage. The four of us flew up to Seattle loaded with lamps, kitchen ware and lots of other accessories for the boat.

We prepared for cruising by having all five fuel tanks filled with diesel fuel. There were two mid-ship tanks, two tanks in the lazarette, and one fifty-gallon day tank, low in the center of the boat, just above the keel. Starting out on our trip, we decided to feed the engines from the day tank. With the fifty gallons available, we felt that it would be three to five hours before it would be necessary to switch tanks.

We cruised along, enjoying the scenery and the new yacht. After about an hour and half of cruising, both engines suddenly stopped. We quickly went to the engine room, but we could see nothing wrong, except that it seemed that the engines had run out of fuel. What neither of us had realized is that a diesel engine draws in more fuel than it burns. This being the case, there is a fuel return line that takes the excess fuel and returns it to another tank. In the case of the day tank, the return fuel is sent to one of the main mid-ship tanks. Since those tanks were already full, the excess fuel had nowhere to go except out the overflow vent pipe. While we had been running, most of our fifty gallons had been pumped overboard. After running the engines dry, each of the fuel injectors on the two six-cylinder engines needed to be primed before the engines would restart.

In the meantime, we were in the channel with a fast-moving current. As we had been assessing our situation, we realized that we would soon be passing under a tall highway bridge. The water in the channel was very deep, too deep to acquire reliable anchoring. We had no choice but to wait and hope that we would not be carried into one of the several concrete bridge supports.

Helplessly we watched as the current would soon place us under the bridge. Luck was with us because, although one of the supports came close, we cleared the bridge with our increased heart rates and blood pressure being the only casualties. Returning to the engine room, one by one we bled the fuel injectors and got both engines restarted. We had learned the hard way how to manage the more complex fuel system installed in that boat.

CHAPTER 27

THE FOLLOWING YEAR DORI AND I were invited by Helga and Mark to join them on the Ponderosa to cruise to Alaska. Since some of us were still working, we had only three weeks to accomplish this lengthy trip.

We met our friends in Friday Harbor, which became our trip starting point. Our initial leg of the trip followed up the east side of the Straits of Georgia. As we prepared to leave a place called Secret Cove, we started the port engine. The starter motor continued to run even after the engine was running. I quickly went into the engine room to look for master switches in order to disconnect the starter from battery power. No switches were visible.

I asked Mark for some wrenches so that I could disconnect the battery cables from the battery. Battery by battery I disconnected cables, but the starter continued to run. By this time, the starter was seriously overheated, and it began to smoke. As I got to the last battery cable, the starter stopped. However, by now the motor insulation had begun to burn. There was no choice but to discharge a fire extinguisher to put out the now visible flames. The fire was extinguished immediately but now the engine room was covered with the white powdered chemicals from the expended extinguisher.

Since the boat was still quite new, we hoped that the boat dealer back in Paulsbo would take some responsibility for helping us with the problem. It turned out that the dealer offered to put a new starter motor on a plane to be delivered the next day. The nearest airport where a land-based airplane could land was in Campbell River across the Straits of Georgia on Vancouver Island.

With our one good engine, we took off for Campbell River. Our trip across was uneventful and we docked waiting for the new starter to arrive. The delay was, however, eating into our precious limited time available for our trip. The new motor didn't arrive until early evening.

Starter motors for diesel engines are quite heavy, especially since the one I needed to install was on the outside of the engine with little space between the diesel engine and the adjacent fuel tank. After a bit of a struggle, I got the new motor mounted, allowing us to continue on our journey.

It was summer, and we were well north so daylight lasted until about 10:00 p.m. So even though we got a late start, we ran for several hours before arriving at our next port. We even successfully took on some rapids that were running strongly due to some inopportune timing of the tides.

On our way north, we stopped at a number of interesting and sometimes desolate places in both British Columbia and southeast Alaska. In British Columbia, we docked one night at an abandoned fish camp. The camp had served as a processing facility for the packing of fish for shipment to Japan. The buildings were open and all the lights were on, even though only one caretaker was living there. We later found out that the lights were on in order to load a water turbine-driven generator located a quarter mile back in the woods. It appeared that the turbine and generator had not been serviced for years.

As the skies darkened, a young fellow landed a pontoon equipped ultra light by the camp and spent the night. His straggly appearance indicated that he had been roughing it. He was wearing a well-worn set of jeans with a sizable split in the seat. It was obvious that he was wearing no underwear beneath.

Farther north we made stops in Ketchikan and Petersburg, Alaska. Fishing is the major industry in Petersburg. Many of the residents there are of Scandinavian descent and run their fishing businesses efficiently and earn good profits. We were surprised to learn that the per capita income in Petersburg is the highest in the U. S. While we were there, we ate at a restaurant south of the town. The grilled halibut steak was the best I had ever eaten.

Still farther north, we stopped in Juneau. Here we rented a car and drove out to view the Mendenhall Glacier. From Juneau, we turned westward heading toward Glacier Bay. Stopping at the ranger station at the entrance to the bay, we were fortunate to get a three-day permit to visit the bay. (Permits are limited in order to protect the natural habitat.) We took a short hike in the area near the ranger station. We were told to be aware of the bears which frequent the area. On our walk we came across a fresh pile of bear scat. Quickly we headed for the boat dock and steered the boat into the very large bay.

As we proceeded deeper into the bay, the water became a semi-opaque turquoise color. This color is caused by fines deposited in the water by the glaciers. We anchored in a spot where we were all alone. Using the dingy we carried, I fished and caught three salmon in short order. The species I caught were called Pinkies because the flesh was pink rather than the more traditional orange color. They weren't as flavorful as most salmon we are used to. We suspected that the silt–laden water may have had something to do with it.

Luck was with us while we were in Glacier Bay since we enjoyed three perfect cloudless days. We were told that such perfect weather is a rarity as there are at least three hundred days per year that are rainy and overcast. As we explored Glacier Bay, we lost track of the many glaciers, small iceburgs and hump-backed whales we sighted. Occasionally, we would hear a loud boom as a glacier would calve into the bay. We were glad that we had anchored far enough away that the wave action from the many tons of ice falling into the water did not bother us.

As we began our trip to the south, we made a stop in Baranoff Hot Springs. Here there was a dock with a small store located above. There were natural hot springs that were led into a bath house with individual galvanized tubs that were continually flushed with hot spring mineral-containing water. A good leisurely soaking in these tubs was a welcome break from the quick showers we had been taking on the boat.

Farther south, we stopped again at Ketchikan and filled the tanks with enough fuel to cover the remaining seven hundred miles of our trip. We had stayed within our three-week schedule, but we wished that we had had more time to linger in some places and to explore the many other great places to be seen in Alaska.

CHAPTER 28

AFTER SEVERAL YEARS OF CRUISING in the northwest with the Ponderosa, Helga and Mark found that the dealer who had sold them the Ponderosa was now carrying a new line of boats that were more stylish and luxurious. The new line was named Vantari. Our friends opted to trade in the Ponderosa for a new 53' Vantari.

One major advantage of the Vantari was that it was powered with Caterpillar diesels, rather than the turbo Fords the Ponderosa had. The Fords were O.K. but somewhat underpowered (225 H.P. each). The Fords had to work hard to push the heavy Ponderosa, causing us to have to add oil and coolant to the engines nearly every night. The Cats were 375 H.P. each, which was a more reasonable amount of power for the size and weight of cruisers of this size.

Once again, we were invited to accompany Helga and Mark on another trip to Alaska. By this time our friends had retired, removing some of the time constraints for this cruise. Starting from near Sidney on Vancouver Island, our friends ran the boat to Ketchikan, Alaska, where we flew in to meet them. Thus, we missed the first seven hundred or more miles of the trip and we could spend more time cruising with them farther north. The new boat was great. We were able to run at a higher cruise speed without overworking the engines. Adding oil and coolant was no longer necessary.

After a long day of cruising we got anxious to make port in Petersburg, Alaska. We put the boat on plane, making eighteen to twenty knots. It wasn't long before the boat got heavy and sluggish. Steam began to come out of the engine room. As I opened the engine room access door, I saw steaming hot water spewing from the rubber exhaust hose for the port engine.

Once again we had to shut one engine down and proceed on one engine. Upon arriving in Petersburg, I was sure that, since there was so much fishing done there, I would be able to find a section of hose to be able to fix our problem. Most boats use rubber hose in the exhaust system. Normally the hot exhaust doesn't burn the rubber because the exhaust is cooled by being mixed with reject cooling water prior to entering the rubber section. What happened with the Vantari is that the water-mixing diffuser allowed the water to enter the rubber hose unevenly, producing hot spots, one of which had burned through.

The replacement hose we needed was eight inches in diameter. Checking with all of the marine supply stores in Petersburg, we found that the largest hose available was six inches in diameter. It was necessary to order some eight inch hose, which had to be delivered from Seattle.

Because of our delay, Mark and I flew back to the bay area in order to take care of some business at the Lab. We left Helga and Dori on board the boat until we returned. Our wives experienced some interesting times watching the loaded fishing boats come in. They off-loaded their catch into large fish processing ships docked in the harbor. These fish-processing boats operated twenty-four hours a day. Our wives also hiked in some areas just outside of the town. They were later told that they had been in areas frequented by bears.

When my friend and I returned, the hose that we needed had arrived at the airport. It was now my job to remove the old hose and replace it with the new piece. It was not an easy job because the hose was thick and not very flexible.

Even though we could now run again, our speed needed to be limited because, with hard running, the hose could burn through again. A newly-designed water diffuser was needed to make a permanent fix. Later, when the boat returned to the Seattle area, new diffusers were installed on both engines, providing the permanent fix that was needed.

Our trip farther north proceeded with no further incidents. In Juneau we considered boating north to Skagway. Skagway is ninety miles north of Juneau, a nearly ten-hour run up and another ten-hour run back. We, therefore, opted to engage an air taxi to take us to Skagway. We rented a car in Skagway and drove to Whitehorse in the Yukon Territory. Our car trip wasn't as interesting as we might have hoped.

Leaving Juneau, we headed west through a channel called Icy Straits. Since we had been to Glacier Bay before, we by-passed it and headed to Sitka. When we arrived in Sitka, we dropped Dori off at the fuel dock in order to look for dock space.

Moving through the marina, we passed by some larger fishing boats facing bow out. The channel was tight. Mark allowed the wind, which was brisk, to carry the boat too close the fishing boats. He tried to steer away, but the stern swung into the bow of one of the fishing boats. There was a sickening crunch as fiberglass was cracked and the rub rail was torn loose on the Vantari. The fishing boat was unscathed. Helga and Mark were terribly distressed at the damage their new boat had experienced. Mark was reluctant to even go back to pick Dori up, so I took over the helm and was able to dock the boat successfully.

After we all got over the unpleasantness of our arrival, we enjoyed our Sitka visit. There were a number of interesting places to explore. There was even a par course in one of the pretty parks up on a hill overlooking the harbor.

After Sitka, we headed south again, covering many of the stops we had made before. Of course, one of them had to be Baranoff Hot Springs, where we once again enjoyed the delightful hot-spring water baths. There was a very nice sixty-two foot Tollycraft docked in front of us. While we were admiring it, the owner and his wife invited us aboard. As we talked to the owner, a seemingly wealthy man, we talked about the size of boats. The owner said that his Tollycraft was as big as he wanted to go. He came up with a statement that we will always remember. He said, “If I can’t run it myself, screw it!”

The rest of the trip was accomplished without incident. Although we have often thought that we would like to, we have never made another small boat cruise to Alaska. Helga and Mark kept the Vantari in the Pacific Northwest for another ten years. During most of these years, Dori and I were not on board. It seemed to us that whenever we had traveled with them there was always trouble. However, each time we had been traveling with them, their boats had been quite new and just needed to have some bugs worked out. In later years, their boating had become stable with no major mishaps occurring.

CHAPTER 29

IN 1980, THE MANAGEMENT OF the weapons program changed. A new associate director for nuclear design was put into place, as was a new associate director for military applications. The associate director for nuclear design chose me to be his deputy. Since the AD had responsibility for both A and B Divisions, and since he had come from A Division, he focused more on the A Division area, while I focused more on the B Division area. The AD and I worked very well together. He and I were both runners. While running at lunch time and on trips that we took together, we were able to spend some quality time talking about critical organizational and personnel issues that were taking place.

The side activity that came as a part of the deputy position was to be a principal Lab advisor for a classified program of cooperation with the French. This program was interesting for its technical content and afforded opportunities for travel to France for periods of a week or two at a time.

As the deputy AD, I continued to maintain a strong interest in Army tactical nuclear weapons. In 1982, I received an appointment to the Army Science Board, a position which I held for a period of six years. Each year the Board took on examination of various Army programs and facilities. One important study that I was involved with was how the U. S. would be able to recognize real signals that eastern-block countries and the Soviet Union were actually planning to invade. Such assessments were always confounded by their regular military exercises, which were hard to distinguish from the real thing. With the Board assignment, I had the opportunity to visit a number of Army facilities. These included West Point, Fort Sill, Fort Belvoir, Fort Rucker, and others. My period of service on the Board was interesting, enjoyable, and educational.

CHAPTER 30

IN THE DISCOVERY BAY COMMUNITY, I took on the civic duty of becoming a Fire Commissioner for the Byron Fire District, which included Discovery Bay. In this position, the commissioners provided guidance and budget sign-off authority for the activities of the local stations. One important input that I made as commissioner was to draft a letter of justification for the acquisition of a new engine for the newly-constructed fire station in Discovery Bay. My letter of justification was successful in that the county granted us the funds with which to procure the new engine.

Commissioner input was very important at that time because Discovery Bay was in the process of growing rapidly, requiring fire and emergency response capability far beyond what had been available before. At times, commissioner interactions with the Fire Chief became somewhat confrontational. The Chief had been in that position for some time. He had some difficulty in accepting the increased needs of a growing community that the commissioners thought were necessary. I served on that commission for a period of eight years and was gratified that I felt that my input had been valuable.

CHAPTER 31

AFTER SPENDING ABOUT TWO YEARS as deputy associate director for nuclear design, I was reassigned to be deputy associate director for military applications in 1982. My job in the new office was to back up the AD in his absence and to focus on guiding and supporting the military requirements representatives residing in the office. These three representatives had different areas of responsibility. One handled strategic offensive systems, another had the tactical nuclear systems, and a third had defensive systems. Each of these areas was quite active at the time.

In about 1984, the Laboratory was heavily involved in the exploration of x-ray lasers. The x-ray lasers being looked at were to be pumped by the explosion of a thermonuclear device. Conceptually, bundles of lasers would surround a centrally located bomb. A strong photon emission from the bomb would stimulate lasing in the surrounding bundles. The laser beams would be emitted before the entire package was vaporized by the exploding bomb. The idea behind such a system was that the collimated x-ray beams emitted from the x-ray lasers would travel through space and impinge on incoming enemy weapons, producing sufficient mechanical damage that the weapons would be neutralized.

Since this entire area was so complex and because it was growing rapidly due to increased funding, I was asked to leave my position as deputy associate director and become deputy program leader for the x-ray laser program activity. In my new position, I was able to make contributions regarding the type and configuration of the source device.

I was granted a top-notch team of senior technical leaders representing engineering, materials science, nuclear testing, and diagnostics. To maintain schedules and keep the program on track, I instituted regular meeting and reporting practices that were held to rigorously.

While I was in that position, there was another change in the overall weapons program management. The person who had been associate director for nuclear design became the associate director with responsibility over the entire weapons program. For a short while the new AD had had a deputy, but that person was called upon to move to the director's office. At that point, I left the x-ray laser program and became principal deputy associate director of the entire weapons program.

Due to some controversy with some of the most senior people at the Laboratory, the associate director that I had been working for decided to resign. A search was immediately begun to find a permanent replacement for the associate director who had resigned. The person who was finally put in that position had been the leader of A Division. I continued to serve as the principal deputy for the new associate director. I remained in that position from 1985 until I retired in the fall of 1991. Incidentally, the AD that I was deputy for is now the Livermore Laboratory Director.

CHAPTER 32

RETURNING TO MY PERSONAL LIFE, we continued our search for a larger boat with better accommodations than we had with our thirty-four foot Fiberform cruiser. We looked at a number of boats in the size range between forty-two to forty-five feet. In this process, we came across a used forty-five foot 1979 Bluewater.

It had two bathrooms and master sleeping accommodations on a lower deck that were extremely comfortable. In appearance, the Bluewater looked somewhat like a houseboat; but, the hull configuration was more of a cruiser type, with sufficient freeboard to allow it to sustain reasonably rough water. It was not, however, a boat which one would regularly use in the open ocean. The Bluewater was powered by two 454 CID Crusader engines. The output of each engine was 325 h.p. We used this boat frequently for yacht-club cruises in the delta and on cruises to San Francisco Bay.

When we got the boat, there was some evidence of wear and tear, since the boat was already five years old. We proceeded with some refurbishments and upgrades. For example, the two sofa hide-a-beds in the main salon were reupholstered. All of the old carpet was replaced.

The wiring under the instrument panel on the flybridge was a mess. Various components had been added in a haphazard, unprofessional way. I installed new instruments and radios and cleaned up all of the messy wiring that had been there. In the engine room, I installed a large bank of heavy-duty deep-cycle batteries, along with a high-capacity battery charger. Many other detailed improvements were made in the engine room and to the plumbing system. After this work, the boat was in top-notch condition. It gave us years of reliable safe operation.

Land n'Sea – 1972

Fiberform Executve – 1979

Bluewater 1984

Camargue – 1994 to Present

CHAPTER 33

IN 1984, OUR OLDEST DAUGHTER Sue was married. We had the ceremony and the reception in our home on December 22nd. The minister we used for the ceremony was a psychologist that the Lab had used for counseling and support of some of its senior managers. On the evening of the wedding, a thick tule fog enveloped the area around Discovery Bay. The fog delayed the arrival of the minister and caused some consternation amongst the guests, who would be leaving the reception in fog and darkness.

The wedding came off without a hitch. Sue and Rich had written their own vows and recited them during the ceremony. The beautiful candle light service took place in front of our hearth that was decorated with an abundance of white poinsettias. The full catered dinner was enjoyed by all. Some of the wedding guests, who were medical students and friends of the Rich, our son-in-law, had been invited to stay over.

Since we had recently acquired the 45' Bluewater, we were able to provide sleeping accommodations for some of them aboard the boat. As people finally departed and began to go to bed, I decided to go make a check aboard the boat to see that everyone had what they needed. As I went down to the lower deck where there were two double berths, I stepped on something soft that emitted a loud grunt. I was rather surprised to see a large dog sleeping on the floor between the two berths. The dog was not a problem, but we were unaware that a dog was a part of the wedding contingent.

A little over a year later, we hosted another wedding at our house. Our son Bill was to be married on June 14, 1986. Since this wedding was planned to take place in the summer, it was decided to have the ceremony and reception outside on our large back deck. To assure better conditions for the ceremony, we had the deck modified by adding a permanent framed-glass windbreak behind the area where we would have the ceremony. A floral arch was constructed to serve as a back drop. Our son-in-law Rich, who sings and plays the guitar, provided a musical prelude before the ceremony. Having had these two weddings at our home, we felt that it had been a great way to bring family and friends together in a more intimate setting.

Pre Thanksgiving Party in 1984

Linda, Sue, and Bill – 1984

Sue and Jackie in 1987 – First Grandchild

AGE 8 WEEKLY BULLETIN WEDNESDAY, MAY 14, 1986

Seeing the Lab

The Hon. Dr. Jay R. Sculley (second from the right), the Asst. Secretary of the Army for Research, Development and Acquisition, visited the Laboratory on May 6. He received a Laboratory overview, briefings on the Free-Electron Laser, Defense Systems and advanced conventional munitions, as well as a tour of the JANUS Combat Simulation Laboratory. At the left are LLNL Associate Director for Defense Systems George Miller and Bill Scanlin, Principal Deputy Associate Director for Defense Systems. Sculley was accompanied by Col. Mike Bissell

Newclipping

CHAPTER 34

IN MAY OF 1987, A group of Discovery Bay Yacht Club members took their boats to the San Francisco Bay for the main purpose of being able to witness the 50th Anniversary festivities planned for the Golden Gate Bridge. All of the Discovery Bay Yacht Club boats docked in slips at the South Beach Marina, located along the Embarcadero south of the Bay Bridge.

The festivities for the anniversary celebration included a pedestrian walk across the bridge. The bridge was closed to vehicle traffic for the day. In the evening there was to be a large fireworks display from the bridge. These activities were well advertised, and it seemed that the city and the influx of people from the surrounding areas were well primed to celebrate and party for the once in a lifetime event.

The pedestrian walk, which most of the Discovery Bay Yacht Club people chose to do, was set to start early in the morning of 24 May 1987. Accordingly, MUNI provided free bus service from just a few blocks from the South Beach Marina to the southern approach to the bridge.

In 1987 we still had our 45' Bluewater cruiser. As guests aboard our boat, we had Mary and Roy Woodruff. Roy had been associate director of the Livermore Weapons Program. I had been Roy's deputy for some years before during the early 1980s. On the morning of 24 May, we awoke very early. As I recall, the sun was just coming up as a number of us walked to the MUNI bus stop. We were surprised that, even at that early hour, the bus was packed and had standing room only. We were dropped off at the bridge approach and saw that the crowd of people starting across the bridge was already very large.

The mood of everyone seemed festive as we began our walk to the north, accompanied by thousands of others. What we expected was that we would walk to the Sausalito side, stay there for a while, and eventually walk back to the San Francisco side. As we got near the center of the bridge, we encountered a mass of people coming from the north. The crowds going in each direction were extremely dense. There was no way either group could move in the direction that they wanted to go. Everything came to a stop.

People farther back in each group had no way of knowing what was happening in the middle. The crowd density continued to increase as those in the back continued to move forward. Loudspeakers on the bridge advised everyone to keep moving. Obviously, that was impossible. All you could do was stand in place, immobilized on a piece of pavement not much larger than the space occupied by your two feet. Some people had brought pets, which had to be lifted up on their owner's shoulders for safety. Bicycles had to be abandoned and were stacked against some of the bridge supports.

The situation was uncomfortable and disturbing, the crowd remained well behaved. If panic had occurred, trampling could have resulted in substantial loss of life. As I stood there thinking about the situation, a little mental arithmetic made me realize that the load on the bridge was probably greater than the bridge design engineers had ever planned for. Even bumper to bumper cars on the bridge would result in less pounds per square foot than people each occupying roughly a square foot of space.

The thought crossed my mind that we could be part of one of the greatest human tragedies ever encountered in modern civilization.

If the bridge were to collapse, a half a million people would have been dumped into the frigid waters below. Fortunately, that didn't happen. But, photographs taken of the bridge, fully loaded with people, showed that the bridge road bed had deflected about eight feet below its normal height. Estimates from that time indicated that the crowd on the bridge had been over one half million people.

After being immobilized for perhaps an hour or more, the crowds from the north and south slowly began to make their way back to their respective starting points. Public transportation was overwhelmed and was basically unavailable. We made our way back to the marina on foot, a distance of at least five miles.

Our walk on the bridge turned out to be one of those life experiences that remains unforgettable.

CHAPTER 35

IN 1985, WHILE SERVING AS principal deputy associate director for the weapons program, I was assigned to participate as a member of the blue ribbon task group reviewing and making recommendations regarding nuclear weapons program management. This group met regularly and always at DOE headquarters in Washington. Nearly every week or two over a period of some months I traveled to and had short stays in D. C.

In 1986, there was an increase in political pressure aimed at curtailing nuclear weapon testing. All of the DOE laboratories continued to make the case that nuclear testing was needed in order to support a safe and reliable nuclear stockpile. Political pressure for cessation was also prevalent in the NATO countries in Europe. To counter this political pressure, it was decided to put together presentations that explained why it was important that full-scale underground testing should be continued.

As a part of this activity, a small group, headed by Frank Gafney and Robert Barker traveled to all of the NATO capitals to explain the U. S. position. I attended and supported these briefings and stood in for Barker while he and Gafney covered another assignment in the

UK. This effort was exceedingly interesting and very rigorous, since we visited a new country almost every day. While this particular effort seemed to be successful, we ultimately entered into a comprehensive test ban in 1991.

On the way to test cessation, there was a period of U. S. – Soviet engagement in test limitation talks that took place in Geneva. These talks, labeled LTBT (Limited Test Ban Talks), were aimed at assuring that both the Soviet Union and the U. S. performed nuclear tests, having yields no greater than 150 kt. Such yields could be verified, using radio chemistry, but using such a technique can reveal sensitive information about each other's test device. Thus, the technique that needed to be used had to be less intrusive. Seismic methods are far less intrusive but are affected by the particular geology of the medium in which the weapon is fired.

These talks went on for a period of months. The U. S. delegation consisted of a mix of people having different areas of expertise. Basically, the U. S. State Department was in charge but used senior Lab managers, who were declared ambassadors, for the duration of these talks. Delegation support consisted of representatives from the State Department, intelligence agencies, the DOE, and Laboratory scientists.

At Livermore three weapons designers were selected to be part of the Livermore contingent. I was selected to be one of the three. We used three because each of us had other duties at the Laboratory that we had to continue to fulfill. The process that we chose to use was to go to Geneva for two weeks, come home for four weeks, and then return again.

While present there, the duty was difficult from the standpoint that the hours were long, often from 8:00 a.m. to 8:00 p.m. Delegation meetings often lasted for hours in a crowded, stuffy, highly secure room. Frustrating to the technical people were the Government lawyers who argued over seemingly trivial aspects of proposed treaty language. A good deal of our work involved preparing proposed language that we would put forth and discuss with our Soviet counterparts.

Fortunately, I was heavily into running at that time. I managed to work in runs either on the lunch hour or in the evening. The U.S. Mission, where the meetings were held was on a hill overlooking Lake Geneva. Thus, it was an easy run down to lake level, but it was tough running back up. These runs were important to me for working off the frustrations that developed from the many long meetings that we had.

Initially, my stays were in a hotel in Geneva. The hotel and the restaurant meals in Geneva were quite expensive. We soon found that housing and food was less expensive and better tasting in France, whose border was quite close by. In the summer time, I often ran across the border in both directions, wearing shorts with no pockets, without a passport or any identification. The border guards seemed to know who I was, and they let me go unchallenged.

The protocols developed by the LTBT talks were ultimately quite successful. The U. S. Senate ratified the protocols by a vote of 98 – 0.

About the time that the nuclear weapon test limitation talks were going on, I had continued to develop my ideas regarding a new way to build nuclear weapons that could have had a dramatic effect on weapon safety and security. I felt that these new designs had the potential for the military to base and store the weapons in a different way. I argued that such basing options could provide much improved survivability of the weapons as well as providing weapons that had dual-purpose capability.

The Laboratory had pulled together a group of retired four-star generals whose purpose was to advise the weapons program on future research and development directions. The chairman of this group of generals was General Andrew Goodpaster, who had been Supreme Allied Commander in Europe and Commandant of West Point.

I put together a briefing for the generals which argued for the new ideas for weapon design. The positive reaction by the generals to what they heard was that they concluded that each of the four-star commanders in Europe should hear it. They made contact with these commanders, who were located in Heidelberg and Frankfurt, Germany and in Naples, Italy. I coordinated visits to each of these commands to coincide with times when I would already be in Europe, in Geneva.

My first visit was scheduled to be with the Commander of the U. S. Army in Europe, located in Heidelberg. I was advised that I would be flown from Geneva to Heidelberg and that I should appear at the Geneva Airport at a prescribed time. I was overwhelmed by the courtesy and respect that I received as I watched a good-sized twin-engine airplane, with the U. S. of America painted on the side, with two pilots, and two flight attendants just for me. When the plane landed in Heidelberg, a large black Mercedes, with armored doors and bullet-proof glass was there to take me from the airport to the General's quarters. I was also impressed by the fact that I was given an hour with the General in a one on one meeting.

My treatment with the Air Force at the Ramstein Air Force base outside of Frankfurt was similarly gracious and respectful. I arrived in the evening and was escorted by several colonels to a dinner in the officer's club. I was then taken to V.I.P. lodging and was shown a very comfortable room. In the room was a privately-labeled bottle of wine with a welcoming note signed by the commanding general. The next day I briefed the general and his staff. Once again, my ideas seemed to be well-received.

For the Navy, I drove a rental car from Rome, where my commercial flight landed, down to the naval headquarters, located in Naples. Much of this command's activity had to do with submarine operations in the Mediterranean. Thus, my proposals were probably somewhat less appropriate for nuclear systems in that area.

I think that pressing for development of these new ideas was unfortunately a little late. Underground nuclear testing ceased in 1991. Because the ideas that I was proposing were a somewhat radical departure from what designs had been, full-scale nuclear testing would have been required to support and certify them.

In view of where the nuclear weapons program is now, had we succeeded in pressing forward with those new ideas, we would have a much easier job in carrying forward with a much smaller, less-hazardous manufacturing complex. Also, arguments supporting stronger safety and security attributes would have been much easier to fulfill.

In June of 1989, I had the opportunity to put forth my ideas for a reconfigured stock pile, which I believed could allow new basing and deployment options. My presentation was made to the NATO Senior Level Weapons Protection Group (SLWPG). This presentation needed to be made at the unclassified level, since the members in attendance were representatives from the various NATO countries, including Germany, Italy, UK, and Greece. Once again, because of the efforts being made at that time, we were perhaps on the brink of having sufficient acceptance that a development program could have been initiated. Unfortunately, it never occurred.

CHAPTER 36

IN 1990 AND 1991, THE Soviet Union began to collapse, and the Berlin Wall came down. Concern had developed that nuclear materials within Russia and its satellites was not being adequately safeguarded. Congressional action, sponsored by Senators Sam Nunn and Richard Lugar, provided funding to allow the U. S. to help the Russians in safeguarding fissionable nuclear materials being harvested from dismantled weapons. I was asked to participate as a member of a group that traveled to Moscow in January of 1991 to discuss specific ways that U. S. technology could be applied to help Russia safeguard its nuclear materials.

In preparing for the trip to Moscow, I had been advised that Russian medical facilities did not have adequate quantities of syringes for giving injections. As a result, syringes were being reused. It, therefore, seemed advisable to carry your own, in the event that an injection might be needed while you were there. My son-in-law is a medical doctor and was able to send me several syringes, which I carried with me.

My itinerary from California included a two-day stop in Washington, D. C. so that the group could discuss its positions with DOE management before proceeding overseas. Our flight from Washington included a stop in Helsinki before heading to Moscow. As our flight from Helsinki to Moscow prepared to land, we were handed immigration and customs cards to fill out. One of the questions on these cards asked whether or not you were carrying drugs or drug-use paraphernalia. The syringes that I was carrying for my personal use would certainly have fallen into that category.

I was very concerned about how I should respond in answering that question. If I responded "yes," I felt uncertain of how the Russian authorities would react to my explanation. If I responded in the negative and the syringes were found in a baggage search upon landing, I knew that I could be in trouble. I was, thus, in a very upsetting dilemma as to what I should do.

As the airplane flew over the airport in preparation for our landing, a miserable, snow-covered icy situation blanketed the airport and all of the parked aircraft. It appeared that there was no activity whatsoever at the airport. Shortly after, the pilot came on the intercom and announced that the Moscow Airport was closed due to the weather conditions and that the flight was returning to Helsinki. I was relieved because there was now going to be a possibility to dispose of the syringes that I was carrying. That evening in the hotel in Helsinki I dropped my package in a hallway trash receptacle.

On this trip, I found that my California clothing was inadequate for the rigors of the weather in Moscow in January. For the two weeks or so that I was there, temperatures ranged from ten below zero to ten above zero. In those weather conditions, ordinary street shoes caused your feet to become very cold after being outside for any length of time.

Conditions in Moscow were rather depressing. On the taxi ride into the city from the airport, trucks and cars had been abandoned on the expressway due to mechanical failures, most likely brought on by the extreme cold and the lack of repair facilities. The Government buildings, in which our joint meetings were held, were in poor repair. Restrooms were typically smelly and had oozing, dripping, plumbing. Heat in the conference room was inadequate. The temperature at the floor level was perhaps in the fifty-degree range. Thus, after sitting around the conference table for many hours, one became uncomfortably cold.

Fortunately, the hotel where many of us stayed was operated by Luthansa. At least the rooms there were clean and comfortable. Many of our meals were taken in a U. S. embassy dining room. It seemed that their menu was based on feeding U. S. Marine security guards, who were accustomed to a diet of more than three thousand calories per day. Thus, our team members were clearly supplied with more than ample food.

Only rarely did we attempt to eat in Russian civilian restaurants. When we did, the food was only of mediocre quality. While I was there, we saw the brand-new MacDonald's Restaurant that had just opened. In spite of prices that were high for the Russian citizens, there were long lines of people waiting to enter, to sample the food that was new to them.

In the later part of my Moscow stay, I caught a nasty cold. By the time I was ready to leave for home, I had become quite anxious to leave the rather depressing situation that I had observed in Moscow. As my flight from Moscow to Frankfurt climbed away in Russian air space, I was very thankful to be heading home.

CHAPTER 37

ALL OF THE WHILE THESE overseas trips were going on there was still the activity associated with the program of cooperation with the French CEA. This program typically involved two to four trips per year to France.

These relatively frequent trips to France resulted in my developing a routine for my air travel. At that time TWA was still a strong and desirable airline. As a result, I channeled my frequent flyer miles to TWA and had purchased a life-time membership in their Ambassador Club. For traveling to France, I would schedule a morning departure from San Francisco on a non-stop to New York, Kennedy. From Kennedy to Paris I would fly the early evening departure which arrived in Paris early the following morning. The Paris flight (TWA 800) was one that I came to know rather well.

On July 17, 1996 Dori and I had been in France on business for a week or two and had taken a hotel room at the airport the night before our departure for home. While dressing the next morning, we turned on the TV to look at the morning news. As we did so, we were stunned as we saw images of fire on water from burning fuel from a crash of TWA 800. We realized that it was the very plane that we would have boarded later that morning for our flight back to New York.

Proceeding to the terminal, we made our way to the Ambassador's Lounge to check in and await our time to board a plane for home. Everyone, and especially the club attendants, was in a very somber mood. We were saddened and very concerned over what had happened to that flight. At the time, there was no explanation of why a plane should have simply blown up after having made a successful take off. A bomb on board seemed to be the only plausible reason to explain what could have happened. Later, as we boarded the substitute aircraft, the passengers were silent as we all recognized further aircraft bombings could be a possibility. The near-miss of tragedy we had experienced was something that was very hard to forget.

CHAPTER 38

RETURNING BRIEFLY TO OUR BOATING activities, our friends, Helga and Mark, who we had boated with in the Pacific Northwest, also owned a condominium in Florida. In the late 1980s, they purchased a 28' cruiser, which they planned to keep in Florida for use during the winter months. Once again we were invited to join them, taking their new boat from Miami to the Bahamas. Dori was still teaching and was on Easter break, which limited the cruising time that we had.

Leaving the inlet at Ft. Lauderdale, we crossed the Gulf Stream and headed for a tour of the closer-in islands. With the smaller boat, things were a little cramped, and water and holding-tank capacity was limited. All of us were experienced boaters, however, so we got along just fine.

We kept moving eastward, stopping at different marinas every night. We enjoyed meals of fresh seafood. Conch soup was a traditional item served at lunch. Although cruising through the Bahama Islands was pleasant, especially in view of the warm clear sea water, the overall experience could not compare with the cruises that we had made in the Northwest.

Due to our limited time available, we soon had to head back the island chain to Florida. As we cruised farther west, the wind came up and the water got very choppy. We had a long way to go so we needed to keep the speed up. At cruising speed, the boat pounded a lot. Dori stood near the stern flexing her legs to absorb the shock of the pounding.

We caught up to a 50' Ocean cruiser going in our direction, also making good time. Falling in behind the bigger boat helped somewhat to smooth our way. As we neared Freeport, located on the western end of the Bahamas, we saw very large waves ahead. We watched as the large Ocean cruiser ahead of us rose and plowed into waves that I would estimate at twelve to fifteen feet tall.

We kept on going and soon encountered the large waves too. It was too much for our smaller boat to take. We needed to turn back. The mere act of making a one hundred and eighty degree turn in such waves is quite hazardous. When the boat is beam to the waves, it can be rolled over.

For a few moments, I observed the period of the waves, planning a quick turn while the boat was in the valley between waves. My strategy worked. I got the boat turned around and was shortly clear of those large waves. We now had a following sea and headed straight for the harbor entrance at Freeport.

Inside the harbor, the water was flat. We docked and as we opened the door to the cabin, we were stunned to see that all of the teak cabinetry had fallen off the walls. Helga was distraught that their new boat had suffered so badly.

Dori and I encouraged our friends to take a walk, perhaps to take a swim and have some lunch. After I looked over the situation, I realized that the cabinets hadn't been mounted too securely in the first place. At a nearby boat yard, we obtained some longer screws. With these screws, we remounted the cabinets and cleaned up the mess. When our friends returned, they were amazed that we had been able to put everything back together so quickly.

Because it was springtime, the big waves and wind were expected to persist for several days. Dori needed to get home so that she could return to her teaching job. Therefore, we needed to catch our flight to California scheduled for the next day. The next morning we got an air taxi, which flew us from Freeport to Ft. Lauderdale where we boarded our non-stop flight home.

In a few days, the water in the Gulf Stream calmed down. Helga and Mark were able to cruise back to Ft. Lauderdale in calm water and with more securely-mounted cabinets in the cabin.

CHAPTER 39

IN 1991, THE LABORATORY WAS offering a special early-out package, designed to encourage early retirements. As I viewed the package being offered, it was tempting to accept it and retire at age fifty-nine. Although the position I was holding at the time was very rewarding and satisfying, the retirement offering seemed to be too good to pass up. Accordingly, I applied for retirement, which was to take place the first of October 1991.

I was so heavily involved at the time that my retirement took place that for the following several years I worked the maximum number of hours allowable under the retirement rules. I continued to support the activities of the weapons program associate directorship as well as retaining my responsibilities for the French program of cooperation.

In late 1991 and early 1992 planning began for our daughter Linda's wedding. A number of different ideas for the reception were pursued. One idea was to have it in an air museum in Santa Monica. The idea that was finally settled upon was to have the reception in the Queen's Ballroom on the Queen Mary, docked in Long Beach, California. The wedding was to take place in the Wayfarer's Chapel, located in Palos Verdes overlooking the Pacific Ocean. Jackie, our oldest grandchild was to serve as flower girl in the wedding party.

The setting in the chapel and the entire ceremony was spectacular. After the ceremony, everyone headed to the Queen Mary for a great get together of family and friends. Grandparents, parents, brothers and sisters of the bride and groom, and assorted grandchildren were all in attendance. Even Wayne's white wolf dog Buddy was smuggled in and lay in the center of the ballroom in the midst of all the activities.

Our son Billy performed yeoman's service that day by flying his wife Alissa, two young children, Tim and Chris, and his two grandmothers down to the wedding and returning them to northern California on the same day. Linda and Wayne had wanted the reception to be a grand party to be enjoyed by all. The reception was everything they could have wished for.

CHAPTER 40

ALTHOUGH I WAS STILL SOMEWHAT active in my work at the Lab, Dori and I began planning to do some boating along the east coast of the U. S. We considered buying a yacht capable of cruising south to the Panama Canal and then heading northward to explore east coast destinations. As we studied that possibility, we realized what a difficult and lengthy trip that would have been. An easier alternative was to purchase a trailerable boat that we could haul to the east coast, do our cruising there, and then tow it back to California.

We surveyed trailerable boats that were as large as possible but small enough that they did not require the display of wide-load signs or require permits and escorts. This meant that the beam could not exceed 8' 6". We finally settled on a 28' Sea Ray cruiser. The one we purchased had actually been in a boat show at Alameda in the fall of 1991. As displayed in the boat show, it was fully equipped and had two 200 h.p. Mercruiser V6 outdrive engines. It was also sold with a heavy-duty triple-axle trailer, making it a good candidate for the long-distance haul that we had planned.

On the day we took delivery, we left Alameda in the afternoon and anticipated that, with the speed of the boat, we would be able to make it back to Discovery Bay before dark. As we left the berth in Alameda, we noticed that the boat needed fuel before we proceeded much farther. We took the boat to a nearby fuel dock and filled the tanks. After fueling, as I am always careful to do with a gasoline boat, I ran the exhaust blowers and raised the engine hatches in order to make sure that there were no gasoline fumes in the engine compartment.

We went down the Oakland Estuary heading for the bay and toward the Bay Bridge. Just before going under the bridge, Dori commented that she thought that she smelled gasoline. I agreed and immediately shut the boat down. I opened the engine room hatch and was shocked at what I observed. Several gallons of raw gasoline were in the bilge. This meant that the least little spark could have ignited the gasoline vapors causing an explosion and fire, which would have destroyed the boat, with the potential of seriously injuring or killing us.

As I examined the engines, looking for the source of the leak, I found that a fuel- hose fitting, coming from the fuel filter, had worked loose, allowing fuel to pour into the bilge as the engines ran. Fortunately, I had brought along a crescent wrench with which I was able to tighten the fitting, stopping the leak. Before getting underway again all of the fuel in the bilge had to be removed. Since it was a brand new boat, I had nothing aboard with which I could accomplish the fuel-removal task. I finally took off my sweat shirt and used it as a mop, wringing it out over the side of the boat. Once the bilge was dry, I ventilated the engine compartment well and was able to get underway again.

The boat was running fine until we were about to pass under the bridge at the Carquinez Straits. At that point, we heard a loud bang and thought that we had hit something in the water. We raised the outdrives to check for signs of a collision, but there was none. We lowered both outdrives and tried to go again. As we put the drives into gear, it was now apparent as to what had happened. The port-side outdrive had failed internally. We, therefore, had only one engine and drive available to us for the remainder of the trip.

With one engine and drive, the boat was only capable of cruising at about twelve knots, rather than the thirty knots that we were able to do when both outdrives were in operation. At the slower speed, we were faced with making the later portion of the journey in darkness. After a long, frustrating, and somewhat dangerous journey, we finally made it home in good shape.

The next day we contacted the Sea Ray dealership from which we had purchased the boat. After explaining the problems that we had encountered, the dealership sent out a service truck with a new outdrive, which was immediately installed.

As a part of the package that we needed to accomplish our east-coast journey, we sold our old 1965 Mercury station wagon and shopped for a new heavy-duty truck. As good as the old station wagon was, we realized that it could be a source of trouble for the journey that we were planning to undertake.

We shopped and test drove pickup truck offerings from Chevrolet, Dodge, and Ford. From our shopping, we concluded that the Ford had the most comfortable cab and seats. We considered diesel power at that time, but diesels in the time frame of 1992 didn't perform as strongly as their gasoline-powered equivalents. Of the diesels, the Dodge was the best, but from other considerations a gasoline-powered Ford was what we chose. What we finally purchased was a Ford F350 with a 460 cu. In. engine with dual rear wheels and the trailer-towing package.

Other preliminary work that I accomplished on the boat prior to leaving was that I installed fuel-flow meters on both engines. These meters told us the gallons per hour of gasoline that we were burning and registered the total amount of fuel burned since the last refueling. I also added the optional so-called fresh water cooling system to each of the engines. With these systems, the coolant circulating through the engine blocks is antifreeze, plus fresh water, rather than having salt water circulate through the systems otherwise.

Installing these systems on the boat was a lot of work. One step involved drilling water intake holes through the bottom of the boat. With two V6 engines in the engine compartment there was little room to do mechanical work. In the process of installing the shut-off valves in the holes that I had drilled in the boat bottom, I found it necessary to drop down head first between the front of the engines and the engine-compartment bulkhead.

As I completed my work, I tried to get back out. After several attempts, using all the strength that I had, I was unable to push myself up and out. I called Dori for help. When she arrived, I was beginning to think that we would have to call 911 to help pull me out. We gave one final try, however. As Dori pulled as hard as she could on my belt, and I pressed upward with my arms, I finally got out.

The route we decided to embark on was to trailer the boat to Ft. Meyer on the west coast of Florida. From there, we would proceed eastward up the Caloosahatchee River, through Lake Okeechobee, and out the east side at Stuart, Florida. From there we planned to use the Inter-coastal Waterway to cover the east coast.

CHAPTER 41

WITH MY WORK ON THE boat and trailer completed, we made several shake-down trips to make sure that everything was operating properly. On an early morning in May of 1992, we pulled away, with boat in tow, from our home in Discovery Bay and headed south on Interstate 5. North of Los Angeles, we joined Interstate 10 and began our long haul to the east.

Some of the hilly territory in the southwest cut into our towing mileage greatly, leaving us with an overall average of only seven miles per gallon for the land portion of our trip. Although we felt as if we would never get across the state of Texas (It took three days.), we eventually managed to cross nine states. Once in Florida, we took Highway 75 south to Ft. Meyers on the state's west coast. This was to be the starting point for the water portion of our adventure.

After launching the boat at Jack's Marine, on the Caloosahatchee River, we were fortunate to be able to rent an empty hangar at nearby Ft. Myers Jet Center in order to stow our truck and trailer for the duration of our trip on the water. The friendly folks at Paige Field even gave us a lift back to the marina where we finished stowing gear, fueling up, and planning the first leg of our cruise.

Heading east on the Caloosahatchee, we passed through three locks to get to Lake Okeechobee. Stories of gators lining the swampy banks were just stories in our experience. We took the southern route around the lake and went through two more locks to enter the St. Lucie River, which would carry us to the coast and the Inter-Coastal Waterway near West Palm Beach.

Our first night aboard the boat was spent in a marina at Manatee Pocket. The name as it turned out was indicative of the next day's long, slow, transit through Manatee zones along the ICW. We did spot one of the manatees so the slow speed zones seemed reasonable. In other parts of the ICW we could comfortably cruise at 3400 rpm, or about 32 mph, which would allow us to put 100 - 150 miles a day under the hull. As we traveled, we spent as much time as we could exploring interesting locations. At Cape Canaveral, we toured the Space Center. On another stop, we visited the beautiful old city of St. Augustine. Further on, we spent a night at St. Simon's Island, Georgia. At Hilton Head, South Carolina, we were inundated by a cloud burst. We spent the night at historic Charleston where we explored the town on foot after the rain had ceased.

The next day, we docked at Portsmouth, Virginia, just in time to take in some of the Sea Wall Fest celebrations going on there. Each day we encountered a new adventure.

By this time, we had settled into a comfortable and efficient routine. Nightly, we stayed in marinas where we could shower. Every three or four days we sought a marina with nearby grocery facilities. At least once a week, we tried to locate a marina with a Laundromat. We found, without exception, friendly and helpful marina folks who often loaned us a car so that we could go shopping.

The Chesapeake Bay provided us our first real navigational challenge. We found large areas of relatively rough water, with widely-separated marker buoys. We encountered a pair of yellow buoys that did not appear on our charts. We, therefore, decided that we needed additional navigational help.

As evening approached, we were plowing through some rough water looking for the mouth of the Potomac River when our cell phone rang. As it turned out, it was our son Bill calling from California to wish Dori a happy birthday. His call helped to break the anxiety we were feeling from the rough water we were encountering. In a short time, we managed to find the entrance to the Potomac River and headed up to Washington, DC, where we spent a few days sightseeing.

We located a marine supply store and purchased a hand-held Garmin G. P. S. From then on, before leaving port we spent time programming way points into the G.P. S. With the aid of the G.P.S., the stress of covering large areas of open water was greatly reduced.

While I was in Washington, DC, I was able to walk to the Forrestal Building, where I worked with some of the people in the Department of Energy Headquarters offices. I also flew back to Livermore for three days of work at the Lab, leaving Dori on board alone.

After leaving Washington DC, we headed back down the Potomac to Solomon's Island, Maryland. On our departure the following morning, we were accompanied by a mass exodus of other boats. It seemed strange to us that they all stopped at the mouth of the Patuxent River. We didn't understand what was going on so we began weaving our way through them. Suddenly, a shot was fired, and all of the boats leaped on plane, going at full throttle. As it turned out, unbeknownst to us, we had driven into the heart of the Solomon's Island Fishing Derby.

We were fortunate that when we stopped in Annapolis, we were able to tour replicas of the Nina, Pinta, and the Santa Maria. These three ships were in port in celebration of the Columbus Quinquennial event. After Annapolis, we continued north on the Chesapeake and were relieved to find the Chesapeake and Delaware Canal and much smoother water. The C & D Canal was a twelve mile long short cut linking the Chesapeake Bay and the Delaware River. It is heavy with commercial traffic and rich in history, being first proposed as long ago as 1661. Later, it was supported by Benjamin Franklin in 1788 and finally opened in 1829.

We cruised south on the Delaware River to the mouth of the Cape May Canal. In the City of Cape May, New Jersey, we met up with some life-long friends. That evening we enjoyed a pig roast at a local yacht club. The next day, we enjoyed touring the historic

Victorian houses that are plentiful in that area. I also used this stop to carry out some routine boat maintenance. I had the boat hauled in order to scrub its bottom and to change the outdrive lubricant.

While we were in Cape May, we rented a car and drove to Valley Forge, Pennsylvania, in order to visit with my cousin Marjorie and her husband Hank. As we stayed in Valley Forge, some of our friends from Dori's church, the church in which we had been married, set up a dinner and get together at the Lulu Country Club. This turned out to be a great event because we had not seen many of those friends for a period of more than twenty years.

Returning to Cape May and preparing for departure, locals warned us that navigating the south Jersey ICW was risky because the channel was silted over in some areas and the marker buoy-system was unreliable.

We, therefore, decided on an alternate course for New York that took us north along the Jersey coastline in the open Atlantic Ocean. As we ran north, we couldn't have asked for smoother water. We made excellent time. Unfortunately, it was so hazy that we could barely see the outline of the big hotels and gambling casinos in Atlantic City.

We spent the night in a marina at Sandy Hook, New Jersey. The next morning, again heading north, we were treated to a spectacular view of the Statue of Liberty just as the sun came from behind a cloud to illuminate it with a beautiful golden light. We stopped the boat for a few moments to savor that outstanding moment.

From there, we cruised up the East River and under the Brooklyn Bridge. Some of the neighborhoods that we passed through toward the northern end of Manhattan Island looked rather inhospitable. Many of the buildings were in poor repair and had broken windows.

To get from the East River to the Hudson River required the opening of the low Spuyten Duyvil railroad bridge. We blew our horn requesting an opening. Nothing happened. We waited and blew our horn several times, but the bridge still did not open. We were concerned that it might be necessary to retrace our path through a somewhat threatening-looking area. We were about to turn around when at 9:00 a.m. sharp the bridge miraculously opened.

Much relieved, we proceeded up the Hudson River, passing by numerous towns with familiar names: Tarrytown, Peekskill, Poughkeepsie, Hide Park, and Catskill. Another highlight was seeing West Point from the water.

We finally docked at Albany, where another boater on the dock offered to drive us to Rockefeller Plaza, which we found to be an architectural masterpiece. We spent the rest of the day touring the historic state capitol, including a visit to the Governor's office (at that time Governor Mario Cuomo).

Leaving Albany, we entered the Erie Canal. During the next three days, we became very proficient dealing with the locking process. We went through thirty-one locks as we proceeded up the Erie Canal and the Oswego Canal. The novelty of locking through wore off quickly.

The locking process was often slow and always involved some tedious work to keep from damaging the boat on the rough concrete and stone canal walls. Fortunately, the surrounding countryside and friendly lock operators made the experience more pleasant. By lock twenty, we had risen over 418 ft. The remaining locks dropped us down 174 ft. to the level of Lake Ontario. We spent the night at Port Oswego.

Our next stop was the historic town of Clayton, New York, where we toured an outstanding antique boat museum. Proceeding towards Canada, we cleared Canadian customs, which we were conveniently able to do, using our cell phone. In Canada, we explored the town of Gananoque, a tourist center for the Thousands Islands area and passed several picturesque islands that had been developed into state parks.

The Bateau Channel led us to the historic town of Kingston, Ontario, which is located at the confluence of the St. Lawrence River, the Rideau Waterway, and Lake Ontario. We spent a pleasant afternoon listening to a concert on the green and touring the Royal Military College and Museum at Ft. Frederick.

Toronto was our next way point. To avoid the open water of Lake Ontario, we chose to pass through the Bay of Quinte and the Murray Canal. Toronto's skyline is imposing and dominated by the 815 ft.

tall CN Tower. With that tower being visible from about twenty-five miles, we hardly needed our G.P.S. to find the city. The harbor was extremely busy with Canadians celebrating our Fourth of July. We chose to dock across the harbor at Center Island Park, using a convenient ferry service to enjoy an evening in the city.

Early the next morning, we crossed Lake Ontario and found the entrance to the Welland Canal by 8:00 a.m. We were dismayed to learn that we were required to have at least three people on board in order to transit the canal. After some delay, we finally found a retired waterway dispatcher, who for $100 (and $80 in Canadian lock fees) would help us through the system. With our hired help on the bow and four bales of hay secured to the railings to prevent damage in the turbulent water, we started through the canal.

Since there is no place for a pleasure boat to tie up overnight, once you start through the canal, you must continue. (The canal operates twenty-four hours a day.) Commercial traffic takes precedence over pleasure craft so we often had long waits. It took us nine hours to make the twenty-six mile trip through the locks, where we were lifted an average of forty-six feet each lock.

Our hired hand, who turned out to be a canal expert, was worth the money. However, by the time we arrived in Port Colbourne, we were hungry, exhausted, and didn't want to see another lock for a long time.

The next day, we faced another challenge, proceeding across Lake Erie. As we headed across, our port engine began overheating, so we had to nurse the boat slowly across the lake to the town of Erie, Pennsylvania. After several hours of fussing with the water pump, back flushing the water lines, and still not solving the problem, we contemplated ending our cruise. After a good overnight rest, I awoke with the realization that I might have reconnected the water hoses incorrectly. Once I made the change, the boat ran fine and we decided that we were ready to push on.

Our next stop was at the resort town of Vermilion, Ohio, about thirty miles west of Cleveland. We docked at Valley Harbor Marina, where the owner offered to loan us his car as long as we needed to shop and do laundry. In addition, he didn't charge us for dockage since we had taken on so much fuel.

The next day we headed for the Detroit River. We viewed the Detroit waterfront as we proceeded to Lake St. Clair. In Lake St. Clair our marine radio announced a Coast Guard warning for all boaters to seek shelter from an impending storm. The storm was due to hit around 11:30 a.m. It was then11:05 a.m. With considerable apprehension, we made our way through rougher water and finally reached the relatively protected waters of the St. Clair River.

The next morning we had the start of a pleasant cruise on Lake Huron. However, as we crossed the mouth of the Saginaw Bay, we were confronted with rough waves which caused us to endure an hour and a half of pounding until we arrived at the Au Sable River. At our next docking, we opened the closet in the cabin and found that the pounding had straightened all of the hooks of the hangers in the locker, which resulted in our clothes being piled on the floor. After cleaning things up and regrouping, we were able to enjoy the lovely protected harbor of Oscoda on the Au Sable River.

Thankfully, the next day we were treated to an easy run to Mackinac Island. At Mackinac Island the only forms of transportation are horse drawn carriages and bicycles. We spent some time viewing the historic hotels and grand summer homes built by lumber barons at the turn of the century.

Our next stretch of water was under the Mackinac Bridge, south through the Straits of Mackinac, and on to Lake Michigan. As we proceeded south on Lake Michigan, we made stops at Frankfort and Ludington. From there we needed to cross westward across Lake Michigan to arrive in Milwaukee. We had fog for this entire crossing, but fortunately, our G.P.S., with our way point set for the entrance to the Milwaukee Harbor, took us straight to the harbor entrance without having seen anything but fog for an interval of two to three hours.

After arriving in Milwaukee, we found that we were just in time to see the annual Great Circus Parade. This parade turned out to be the best that we had ever seen. Large antique circus wagons were pulled by a variety of animals, such as magnificent Clydesdales, and in one case even by camels.

Our next destination after leaving Milwaukee was Chicago. The run from Milwaukee to Chicago was more exciting than we would have liked. Throughout the trip the waves were three to four feet high and closely spaced. They came from behind, resulting in a following-sea condition. We alternated between surfing or getting caught in a trough, causing us to fall off plane. Two thirds of the way there the Coast Guard announced on our radio a small craft advisory. We were relieved to reach the Chicago Harbor, marking the end of our rough water treks on the Great Lakes.

Passing through the Chicago Lock from Lake Michigan to the Chicago River, led us in to down-town Chicago. Fortunately, our boat was low enough that it was not necessary to raise the numerous roadway bridges across the river. That evening we stopped at the Marina City Marina, located beneath one of the twin Marina Towers buildings.

That night, docked beneath the tower, a severe thunder storm passed through the Chicago area. There was even a tornado watch. As lightening flashed and thunder rumbled, we were thankful to be snug beneath the tall building above. The next day, we took some time to explore the Merchandise Mart, the Chicago Commodities Exchange, and went up to the top of the Sears Tower.

Our departure the next day put us into rain and limited our speed due to heavy commercial barge traffic and several locks. As we continued southward, we traveled on the Chicago Sanitary & Ship Canal, the Des Plaines River, and the Illinois River. By early evening we found a nice marina at Seneca, Illinois, where we spent a quiet night. The rain continued the following day as we continued on the Illinois River, encountering several more locks. As the day wore on, two marinas that we might have wanted to stop at were closed, forcing us to push onward. We thus spent a long, rather miserable day, in foul-weather gear, having traveled 246 miles before we tied up at Pere Marquette State Park Marina.

The next day, after traveling seven miles south of Pere Marquette we joined the Mississippi River. The Mississippi was intimidating.

Because the current is so swift in the St. Louis area, pleasure-boat facilities are not plentiful. North of St. Louis we located a marina in St. Charles, Missouri. There we rented a slip because it was necessary for us to leave the boat for an interval of about two weeks. This interlude was necessary because I needed to attend a series of technical meetings at both Los Alamos and Sandia. During that same period of time, Dori went back to our home in Discovery Bay.

Leaving St. Charles, we continued farther south on the Mississippi. Due to storms farther north, the river was full of debris. Some of the logs that we saw were large enough to sink a boat. At time, the debris was so thick that we had to stop to remove small branches and other material from the outdrives before we could continue. We had been warned to keep clear of whirlpools in the water, which typically mark logs stuck in the river bottom or channel buoys pulled under the water surface by the strong current.

Fuel and dockage on the Mississippi, is very limited, with only two fuel stops between mile markers 250 and 0. Both of these stops are simply metal barges that you tie up to in order to do refueling. We spent the night at one of these fuel barges at Cape Girardeau.

From that fuel barge, we cruised an additional 52 miles south before reaching the Ohio River, near Cairo, Illinois. We only went 31 miles on the Ohio River before turning off on the Cumberland River, where we enjoyed the first clean calm water that we had seen in weeks.

The next day we cruised through a boater's paradise through the Barclay Canal, which connects Barclay Lake with the Kentucky Lakes, which are on the Tennessee River. In that area, the water was warm, and there were plenty of inviting places to anchor behind islands and in protected coves.

The Tennessee River led us to the upper Tombigbee Waterway. In that area, we saw lots of small boat traffic, launch ramps, and other marine services. However, south of Demopolis there were only two places on the waterway to get fuel. Small boat traffic in this region was virtually non-existent, and there was very little visible civilization. On the Tombigbee, we passed through eleven locks. Several of these locks dropped us 81 feet. As we cleared the Coffeeville Lock, the last of a total of sixty-two, we celebrated.

On the lower Tombigbee, we went to the last fuel stop below Demopolis and spent the night at a small facility called Lady's Landing. It was quite isolated and we were the only boat tied up for that night. It was a colorful spot because goats were allowed to roam in order to keep the grass and underbrush trimmed.

Our boating portion of the trip terminated in Mobile, Alabama. Although we could have taken off across the Gulf of Mexico in order to return to Ft. Myers, we felt that we had had enough boating for the time being. We left the Sea Ray at Dog Creek Marina in Mobile and chose to drive a rental car from Mobile to Ft. Myer in order to retrieve our Ford truck and boat trailer for the long ride back home.

During the tow back to our northern California home, we reflected on the adventure that we had had. We were pleased to realize that we had cruised 4,000 miles all in unfamiliar waters without even dinging a prop. Our entire trip had spanned three months, with thirty-eight days having been spent on the water.

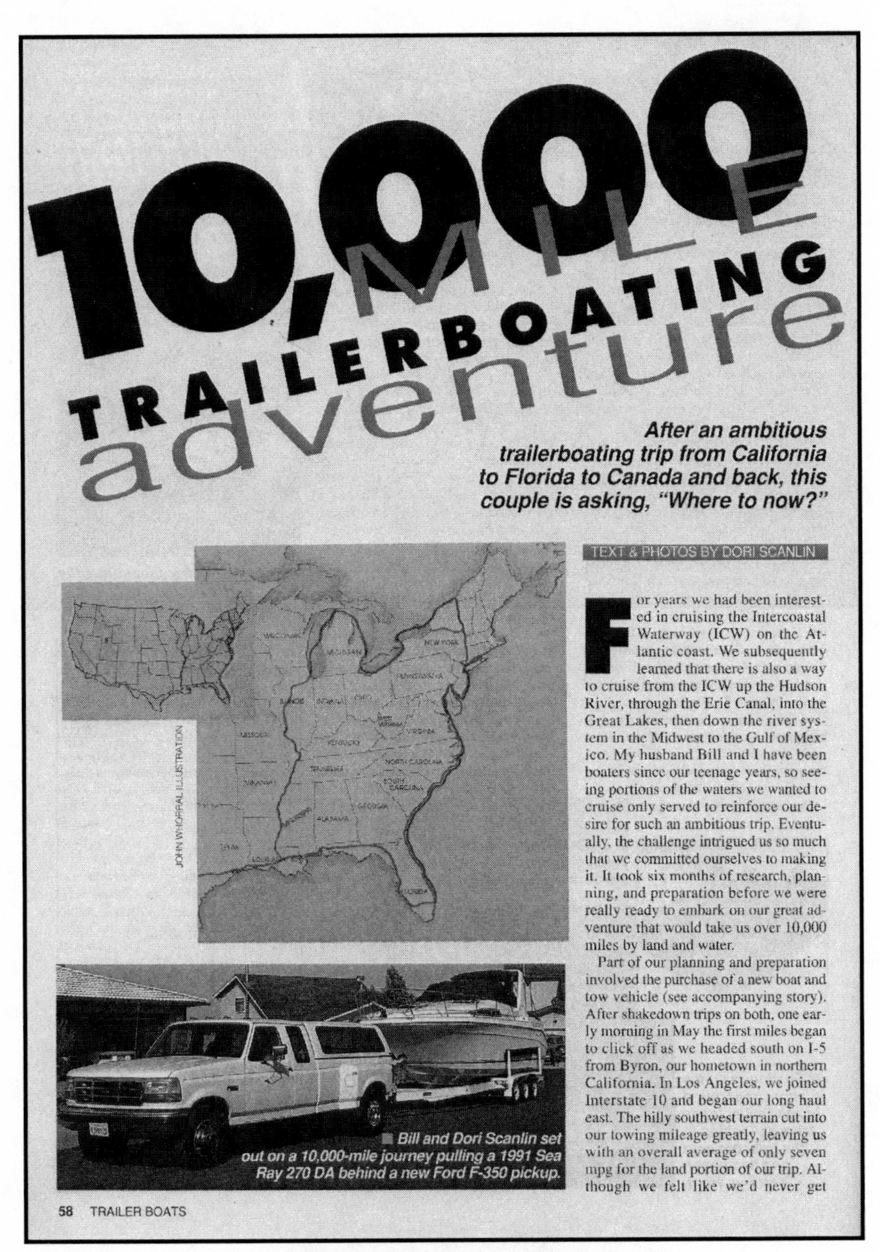

10,000 MILE TRAILERBOATING adventure

After an ambitious trailerboating trip from California to Florida to Canada and back, this couple is asking, "Where to now?"

TEXT & PHOTOS BY DORI SCANLIN

For years we had been interested in cruising the Intercoastal Waterway (ICW) on the Atlantic coast. We subsequently learned that there is also a way to cruise from the ICW up the Hudson River, through the Erie Canal, into the Great Lakes, then down the river system in the Midwest to the Gulf of Mexico. My husband Bill and I have been boaters since our teenage years, so seeing portions of the waters we wanted to cruise only served to reinforce our desire for such an ambitious trip. Eventually, the challenge intrigued us so much that we committed ourselves to making it. It took six months of research, planning, and preparation before we were really ready to embark on our great adventure that would take us over 10,000 miles by land and water.

Part of our planning and preparation involved the purchase of a new boat and tow vehicle (see accompanying story). After shakedown trips on both, one early morning in May the first miles began to click off as we headed south on I-5 from Byron, our hometown in northern California. In Los Angeles, we joined Interstate 10 and began our long haul east. The hilly southwest terrain cut into our towing mileage greatly, leaving us with an overall average of only seven mpg for the land portion of our trip. Although we felt like we'd never get

■ Bill and Dori Scanlin set out on a 10,000-mile journey pulling a 1991 Sea Ray 270 DA behind a new Ford F-350 pickup.

58 TRAILER BOATS

10,000 Mile Adventure 1992

Traversing the Erie Canal in 1992

Philadelphia Area in 1992

CHAPTER 42

HAVING ACCOMPLISHED OUR DREAM TRIP of circumnavigating the eastern U. S., we returned home with the situation where we now owned two boats. The twenty-eight foot Sea Ray was kept on a large hydro-hoist on a dock behind the house. We still had our forty-five foot Bluewater that we had had since 1984. As we did yacht club cruising and other cruising around the delta, there was always the dilemma as to which boat to take. For comfort considerations, we almost always opted for taking the Bluewater. As a result, the Sea Ray was getting very little use and was losing value as it simply kept getting older. Consequently, we decided to sell it.

With that transaction completed, we began thinking once again about obtaining a more capable ocean-going cruiser. One of our friends in Discovery Bay had a forty-two foot Krogen trawler. Krogens are heavy, very seaworthy, boats, but being a displacement hull design, they are incapable of cruising much faster than eight and half to nine knots. We were only mildly interested in a boat of that type.

When the forty-five foot Bayliner, pilothouse cruiser came on the market in 1986, we were enamored with that design. In the spring of 1994, Dori and I drove to Seattle to look at boats that were on the market there. The Bayliner dealer in Seattle had just received a delivery of the first of the forty-seven foot pilothouse cruisers. The forty-seven had a number of improvements compared to the forty-five which we had liked so much over the years. After further consideration, we decided to investigate buying a forty-seven foot Bayliner from the bay area dealer located at Oyster Point.

Our friend, Howard, who had the forty-two foot Krogen had spoken to us about a seminar that was to be given by the owner and designer of Krogen Yachts (Jim Krogen). The Krogen dealer, located in Alameda, had several used forty-twos available to look at. Once again, it was our view that, although the Krogen is a very seaworthy boat, it was somewhat lacking in space, and the sleeping accommodations in the master stateroom were, in our view, inadequate.

Another used boat that the dealer had in his inventory was a forty-eight foot Camargue. The boat had recently been operated by a professional captain. The captain had been tasked to prepare the boat for a trip through the Panama Canal. Much of the equipment needed for such a trip had already been installed.

In addition to the more normal equipment items, such as VHF radio, depth sounder, and radar, it also had a water maker, a single side-band radio (which enabled two-way conversations over very long distances), a weather FAX and six hundred feet of anchor chain. Included in the package was a 3,000 p.s.i. air compressor for filling scuba tanks and a fourteen and a half foot center-console hard-bottom inflatable dinghy. The dinghy was powered by a 30 h.p. electric-start outboard. The dinghy itself was even equipped with its own depth sounder.

We made a list of all the extra equipment that was available and went home to spend a few days thinking about the trade-offs of purchasing a used Camargue versus the new forty-seven foot Bayliner. In our minds, the Bayliner was still our dream boat. We decided to drive over to Oyster Point to make the purchase. However, we decided to make one more last inspection of the Camargue before proceeding across the bay.

Once again, we reviewed the list of all of the items of equipment that the Camargue had. As I examined the construction of the Camargue in more detail, I realized what a high-quality boat it was. At this point, we felt compelled to reconsider our decision. We decided to place an offer to purchase the Camargue. Because the owner of the Camargue was in ill health, he decided to accept the offer. Thus, we became owners of two boats, both the Camargue and the Bluewater.

Before the deal on the Camargue was closed, we had the boat surveyed in an excellent boat yard in Alameda. The survey turned up only very minor defects, which were immediately taken care of. After all work had been completed, we boarded the Camargue in the late afternoon, with the intent of running it back to Discovery Bay. It was low on fuel so we began hunting for a fuel dock that would still be open. We were unable to find one. We finally headed across the bay, hoping to find an empty slip at the South Beach Marina. Fortunately, we found one, tied up, and spent our first night aboard.

The next morning we filled our tanks in Sausalito and headed on home. In the meantime, we had placed the Bluewater up for sale. We were fortunate in that, on the same day we brought the Camargue home, our buyer took the Bluewater from our home to his home, which was also in Discovery Bay. We were sad to see the Bluewater leave since for the past ten years we had had many good times on it.

After the Bluewater had departed, we prepared to bring the Camargue into the slip on the inside of our U-shaped dock. As I slowly brought the boat forward to enter the slip, I was greeted with a rather unhappy surprise. The Camargue would simply not fit. The dock needed to be only six inches wider, but it might as well have been a mile.

Temporarily, we were able to tie the boat up on the outside of the U-dock until we could get a contractor to split the dock and make the slip wider. In doing this, it was also necessary to remove one of the three pilings and install a new one, allowing one of the dock fingers to be moved apart by about two feet.

CHAPTER 43

WITH THE ACQUISITION OF THIS capable ocean-going boat, we began preparing for an extended trip to Mexico. For such a trip, the boat had to be absolutely reliable. It was necessary to acquire spare parts and components for those items which would be most-likely to wear or fail on a long trip. I spent the next year rebuilding pumps and changing cooling hoses and V belts. I overhauled the vacu-flush toilets and added a new VHF radio and other instrumentation. The instruments added were fuel-flow meters for each engine, a digital depth sounder, speed log, and a wind velocity meter.

An extremely useful addition was a chart plotter, linked to a G.P.S., which was already on board. Cartridges installed in the chart plotter showed detailed maps of the regions where we planned to cruise. A G.P.S. provides the information that allows the chart plotter to show the position of the boat relative to the chart being displayed. The chart plotter is probably the most important component in providing navigational confidence during long-distance cruising. However, electronics can fail. Therefore, we prepared ourselves with many paper backup charts.

I had some problems with the anchor windlass, which I removed and returned to the manufacturer for overhaul. I needed to assure myself that our anchoring system would be up to the rigors of holding us securely in strong winds. Accordingly, the forty-five pound plow anchor was replaced with a sixty- pound CQR. The plow anchor would now become a spare. One very important change that we made to the Camargue was to add a solid fiberglass cover over the flybridge. New canvas and isinglass windows made the flybridge completely weather tight. Having made this very important change, the flybridge became in effect our pilothouse.

I have owned scuba diving equipment for many years. Although I have never taken a formal scuba-diving course, I have used the equipment a number of times and would in an emergency be prepared to dive. The dive regulator that I had I had purchased in 1959. I felt that it might no longer be safe. I, therefore, purchased a whole new set of diving equipment, which I carried on board. Tank mounts were constructed and two air tanks were stowed in the lazerette.

To test our preparedness for open-ocean cruising, we took a trip to Monterey. On the return trip we intentionally chose a day where the seas were quite rough. Prior to leaving, we thought that we had properly secured everything. The dining- room table, which had a heavy glass top insert, was hooked to the cabin wall. After the encounter with the rough water, we were surprised to find that the table had become unhooked and had in fact flipped upside down.

Fortunately, the table and its glass insert were not damaged by the encounter. I decided that a more secure permanent fix had to be put in place before we left on our long journey. The problem was fixed by drilling holes from below through the floor and running bolts into tapped holes that I had made in a thick sheet of plastic secured to the table base. In literally thousands of miles of running in the ocean, the bolted down table has never moved. When embarking on any ocean excursion, we spend nearly an hour putting away all loose items and securing any moveable furniture with heavy bungee cords.

Our Monterey trip had uncovered the weakness in securing cabin furniture. But, we also found that, with the canvas bimini top over the flybridge, we were unable to pilot the boat from the flybridge without getting drenched. In that instance, we had to retreat and drive the boat from down below. This finding made us feel that it was necessary to put in place the hardtop and enclosure over the flybridge as we described earlier.

In the fall of 1995, we felt that we had done everything necessary to make the boat ready for the nearly 2,000 mile cruise down along the west coast of Mexico. Marge and Howard, our friends with the forty-two foot Krogen planned to accompany us. Our first step involved getting the boat from Discovery to San Francisco Bay. A day or so later we embarked on our trip down the coast, with the first stop being Monterey. We left South Beach Marina where we had stayed overnight before dawn. While I was putting lines and fenders away, Dori was piloting the boat. She was surprised to see a man in a raft shouting and waving frantically at her. She heard splashing and was astonished to see over a dozen people swimming not far from the boat. Dori immediately put the transmissions in neutral and waited for the swimmers to go by. Apparently the swimmers were on a San Francisco to Alcatraz swim.

Getting underway again, we headed out the Golden Gate, following the coastline down to Monterey, about one hundred miles south. When we got to Monterey, there were no slips available. We, therefore, tied up at the dinghy dock and spent the night. In the morning, as we prepared to leave, some sea lions were lying on the dock. As Dori went forward to untie the bow line, a large sea lion that was lying near the cleat hissed at her. Dori called to me for help. I aggressively walked forward, and the sea lion, seeing me coming, slid off the dock and into the water.

We were now able to head toward our second destination which was Morro Bay. As we approached Morro Bay, we encountered a very thick fog bank. With our radar and our G.P.S., we moved slowly through the fog into the harbor entrance. Very shortly, we broke into the clear with bright sunshine over the harbor. On this part of the trip, our friends Howard and Marge, who owned a forty-two foot Krogen, were accompanying us. For safety and companionship we had planned to buddy boat with them all the way to Mexico. Leaving Morro Bay, Howard and Marge received a phone call that informed them that Marge's mother was seriously ill. As a result, they were forced to abort the trip, and they returned to Discovery Bay.

Our next stop was Santa Barbara. The marina there is quite large and heavily occupied. We were fortunate, however, to be able to tie up in an empty slip. As we walked up into the down town area, we were disappointed in its condition. We had been there before. This time it seemed somewhat dirty and run down. Leaving Santa Barbara, we rounded Point Conception. Entering the Santa Barbara channel, a series of oil-drilling platforms are visible as they sit a mile or two off shore. In this area, while boating near Goleta there was a presence of oil in the water. This oil is due to natural seepage caused by pressure pushing the oil up from deep below. It seems to make sense that additional oil drilling and pumping in the area would help to alleviate the problem.

Our next stop was in the Channel Islands Marina. This stop was completely uneventful. Next, we made the long run off the coast of Palos Verdes and down town Los Angeles. Nearing Long Beach, we could see the Queen Mary sitting in its permanent moorage there. Our stop in Long Beach included a visit with our daughter Linda and her husband Wayne.

After spending several days with them, we decided to take them and their part- wolf Buddy on the long run to San Diego. As we all boarded our boat, we had to lift the one hundred plus pound Buddy up the ladder so that he could ride with us on the flybridge. The ride was fairly calm and we all had a good time.

We arrived in San Diego in early October. Our insurance policy restricted our departure until early November, since that is normally the end of the hurricane season in Mexico. We, therefore, had a few weeks to spend exploring the San Diego area and making final preparations for the long voyage south.

We put on board more supplies. We located boxed milk, which did not require refrigeration. We also stocked up on toilet paper, paper towels, and loaded our separate freezer with all kinds of meat. It turned out that the boat was so well-supplied with paper products that we are still using the paper products that we put on the boat over fifteen years ago.

The West Marine Store in San Diego provided a course on advanced first aid. This course was taught by a medical doctor. As a result, he was able to provide us with prescriptions that we could fill in order to enhance our significant first aid kit. While we were there, we ran into Bajahaha people (a group of mainly sailboaters) also heading south. As our departure approached, our friend, Howard, flew to San Diego in order to join us for our trip to Cabo San Lucas.

Finally, on October 31st, we and the Bajahaha group left the harbor in San Diego. Our boat ran well-ahead of the group of sail boaters. In the northern part of the Baja coast, we started anchoring for over-night stops. These stops were relatively uneventful. However, at the northern end of Cedros Island, the anchoring was difficult because the bottom dropped off so steeply. When the bow was hooked, the anchor off the stern was in such deep water that the stern anchor did not have sufficient scope. We worried all night about the possibility of the stern anchor breaking loose. As it was oriented, the boat was facing with its beam towards the waves. The boat rocked all night. Furthermore sea lions made lots of noise snorting and breathing heavily. We didn't sleep well at all. It seemed as though it was not worth stopping.

A bit farther south we entered Turtle Bay in order to add some more fuel. At that time, Turtle Bay was the only reasonable fueling station down the entire western coast of the Baja Peninsula. The fuel dock at Turtle Bay was on pilings ten or twelve above the boat's deck level. Rumors had persisted about the possibility of cheating going on at this location. We got Howard up on the dock so that we could make sure that the meter on the fuel had been properly set back to zero. The fuel economy of the boat had been very good. As a result, I only needed to load one hundred gallons at Turtle Bay in order to assure ourselves that we would have a good factor of safety in getting ourselves down to Cabo.

At that time, fuel had to be paid for in cash. When fueling was completed, you paid by putting money into a can, which was dropped down to your boat. As we were doing this part of the operation, one of our twenty dollar bills landed in the water. The Mexican kids on the dock practically went wild with the anticipation of grabbing the money. Fortunately, an American in a dinghy was close to the bill and retrieved it for us. Inside this protected harbor, we spent a quiet overnight, with no serious rolling.

Our next stop was at Magdalena Bay, which was roughly two hundred miles farther south. No good anchorages on that portion of the trip were available so we had to run all night. During the night, it was overcast so there was no moonlight. Then we ran into heavy rain. With the rain and the black night, we couldn't see anything. By radio we communicated with the Bajahaha Group and told them what we had run into.

We finally entered Magdalena Bay and turned to the north, anchoring near the beach and a little village that was located there. We dinghied to shore and talked to some the Mexicans and their kids. Howard used his video camera to take pictures of the children. They were amazed to see themselves on video replays that Howard made.

While we were in Mag Bay, we ran into some young Canadians who were aboard a small sail boat. They needed some fuel in order to get to Cabo. My boat has a built-in transfer system so I was able to fill a couple of cans for them. We hoped that it would be enough for them to make the rest of the trip.

On our way from Magdalena Bay to Cabo San Lucas, we had to make another overnight run. There were no harbors at all along this part of the Baja coastline. After running all night, we arrived in Cabo the next morning. It was hot and humid and we were very tired. Some Mexicans wanted to make money washing out boat. We told them to go ahead, but they were careless as they toweled off the water. They pushed in all of the screens on our salon windows.

Dori had been doing wash in coin-operated machines. When she returned and saw what had happened, she was in tears. We had just replaced all the screens before we left and it had been a lot of work.

After a couple of days, we got some rest and got the boat back in shape. Howard got on a flight from Cabo to San Francisco and went home. We explored Cabo on foot and generally were not impressed. Away from the main street, the streets were unpaved and dusty. Prices of food and most commodities seemed expensive. It almost seemed as though it was an extension of parts of Los Angeles.

One of our major destinations was Mazatlan. The crossing

between Cabo San Lucas and Mazatlan is lengthy, about two hundred miles, with the potential of having to deal with a beam sea. The entire length of the Sea of Cortez is to your port, giving wave action from the north lots of distance to develop. Since our crossing would take about twenty hours, we left Cabo in the late morning so that we would arrive in Mazatlan early the next day.

The cruise during the afternoon and evening was reasonably smooth. As night wore on, the beam sea developed. We were tempted to steer north a bit in order to ease the rolling motion, but doing so would have made our trip longer. We elected to hold our course and tough it out.

About 2:00 a.m., we noticed that we were being followed by a large white bird. We believe that it may have been an albatross. It was gliding on the upper wind wave that our boat was creating. We wondered how a bird could survive since there was no land within a hundred miles in any direction.

When we arrived in Mazatlan, we elected to stay at the El Cid Marina. Fortunately, they had a slip for us, which we decided to rent for a month. Services there were great. We had the use of the swimming pool, the restaurant and on-shore restrooms that were well-maintained and clean. The marina and adjacent condominiums were serviced by small open taxis called pulmonias. The taxis were free going to the main El Cid Hotel a couple of miles to the south. For a nominal fare, you could use them to go all the way into the center of town.

We grew to like Mazatlan very much. It is less dependent on tourism. It has enough business and industry that it could survive without tourists. Our situation in the marina was comfortable. Since we used the on-shore restroom facilities almost all of the time, our boat's holding tank was good for weeks without having to dump. Around the town, we got to know many good restaurants which offered excellent food at good prices.

While we were there, I had to take a business trip back to the U. S. Dori stayed on the boat alone. One night she was treated to a fireworks display over the marina. It might have been pretty, but it was not a good idea. Some of the hot particles drifted down and landed on our boat. A number of tiny black pits are now present in our gelcoat. We complained to the harbor master who finally half-satisfied our concern by giving us a substantial discount on our slip rental.

While we were there, we had some friends from Discovery Bay who were cruising on the Crystal Symphony. When the cruise stopped in Mazatlan, they invited us to join them for lunch on the cruise ship. After lunch, we invited them back to our marina so that we could take them for a boat ride along the Mazatlan water front.

As we left the marina and turned south, we were chased by a small boat that had also come out of the marina. As the boat overtook us, we noticed that two security guards from our marina were waving wildly at us. Apparently, an express package had just been delivered to the harbor office addressed to me. They thought that we were leaving to go to another area and wanted to make sure that the package got delivered. They didn't realize that we were coming right back. The package was important, however, because it contained airline tickets sent by the Lab so that I could take another business trip. Our friends were impressed by this very special delivery that we had received.

In late November, our daughter, Linda, came to visit us. She stayed for several days over Thanksgiving. We had fun taking our dinghy out to three small islands situated off the coast. Around these islands there was good snorkeling, which we enjoyed several times while Linda was there.

When our month's docking at Mazatlan was up, it was time to move on to Puerto Vallarta. From Mazatlan to Puerto Vallarta is a long two hundred plus mile run. We decided to break up the trip by making an overnight stop at Isla Isabella. Isla Isabella is a small uninhabited island about thirty miles off the western coast of Mexico. A cruising book we had described this island as a good place to anchor. The book gave information on how and where to set your anchor.

Arriving at Isla Isabella, we followed the book's instructions for anchoring carefully. A large white guano-covered rock was one landmark we used for locating ourselves. At the anchoring point recommended, we were in thirty feet of water, not far from the beach. The anchor went down and seemed to hook up properly. We spent a quiet night all alone thirty miles off the coast.

In the morning, we woke early, planning to reach Puerto Vallarta by early afternoon. After a quick breakfast at about 6:00 a.m., I went to the bow to pull up the anchor. The windless pulled normally, pulling the boat forward until the chain was straight down. At that point, the windless stalled. Starting the engines, I moved the boat forward and backward a number of times, but I could not free the anchor.

I decided that I was going to have to dive to see why the anchor had become so fouled. Since the sun had not yet come up, we decided to go back to bed and wait for full daylight. About 8:00 a.m., I got up and put on all of my diving gear. Rolling off the swim platform, I swam forward to the anchor chain.

Following the chain, down I went. The water was quite clear. As I neared the bottom, it became obvious as to why the anchor would not release. The wind must have shifted several times during the night causing the chain to wrap itself around a large overhanging rock. Surveying the situation, I realized that swimming or walking, carrying the anchor with thirty to fifty feet of chain, circling several times around the eight-foot diameter rock was nearly impossible.

What I needed to do was to unfasten the clevis that attached the chain to the anchor. With no anchor attached, I thought that I should be able to unwrap the chain and use our windless to pull it up. I also hoped that I would be able to attach a nylon line and float to the anchor and pull it up by hand later.

I surfaced and explained to Dori what I planned to do. Since we were close to the beach, she needed to have the engines running and be prepared to move away from the beach as soon as the anchor chain was free and back on board. All the while, she had to be careful not to run over me as I surfaced.

With our action plan clear in our minds, I took the tools and float that I needed and dove back down. At the anchor, I cut the safety wire and used a wrench to undo the clevis. I attached the float and line to the anchor. Letting go, the float took the line and popped to the surface. As I expected, with the anchor removed, I was able to walk the chain around the rock several times until it was free. I immediately went to the surface and signaled Dori to pull in the chain.

As soon as the chain was all in, Dori quickly moved to the helm and put the engines in gear, moving the boat away from the beach, all the while staying away from me while I paddled nearby. Back on the boat, I put on gloves and was able to hoist the heavy anchor aboard. Although we were successfully able to extricate ourselves from this difficult situation, we planned to avoid Isla Isabella on any future northerly or southerly excursions.

In Puerto Vallarta we rented a slip in a good location within the well-sheltered marina. When we arrived, the harbor master and several helpers were on the dock to help us dock and tie up. In some respects, moorage in Puerto Vallarta wasn't as good as we had in Mazatlan.

First, the monthly slip rental rate in Puerto Vallarta was nearly twice as expensive as that in Mazatlan. Second, the on shore restroom facilities were not as convenient and were not as clean and modern as those in Mazatlan. One good feature, however, was that the marina had a small boat with pumping equipment that came around weekly. With this boat, the holding tank could be pumped out with a nominal fee of seven dollars. We used this convenient service regularly.

At the time we stayed in Puerto Vallarta, it had only been a short time since the city had experienced the effects of a fairly large earthquake. One hotel (the Vidafel) near the marina was so badly damaged that it had to be destroyed. Around the periphery of the marina there were condominium buildings that exhibited cracks, fallen stucco, and fractured concrete caused by the quake. While we were there, much of this damage was still under repair.

On our dock, we met another couple who had been cruising from San Diego with a forty-eight foot Taiwan-built sedan model. The owner (Alfred) who was four or five years older than me had had a varied life experience. At one time, he had run a fishing-boat charter business in Florida. Most recently he and his wife (Elizabeth) had run a photography studio in Arizona. Most of his photographic work was doing weddings and other large social functions. Elizabeth was a Philippina, a good bit younger than Alfred.

One activity that Alfred and Elizabeth shared with us was taking some formal Spanish instruction. We hired a teacher, who came to our boats and would spend a couple of hours several days a week working on vocabulary, grammar, and pronunciation. At times, he would give us written assignment by having us write a paper in Spanish on a subject that he selected. The following day, he would orally critique all of our papers.

At Christmas, our son Bill, his wife Alissa, and their two young sons, Tim age five, and Chris age three, at that time, came to visit us on our boat in the marina. There was a small water-slide park and pool near the marina which the boys enjoyed. We used our fourteen foot dinghy to take everyone south of Puerto Vallarta to visit and snorkel at the rock formations called Los Arcos because one formation does in fact form an arch. As Christmas approached we bought a good-sized Christmas tree and put it up on the back deck.

On the night before Christmas Eve, we heard that there was to be a parade of dinghies roving through the harbor. Most of the dinghies were from sail boats that had participated in the Bajahaha. Singing carols as the string of dinghies roamed through the marina was to be part of the fun. As it turned out my dinghy was the largest and most powerful of all of the thirty or so that were to participate. I was appointed to lead the pack.

Getting underway after dark, I had all thirty dinghies, containing about a hundred people tied behind my boat. It was quite a chore keeping the dinghies aligned and staying on course as we weaved in and out of the many fingers of slips in the marina.

On Christmas morning, we hung a piñata that we had purchased previously in a marina-front property not yet developed. The boys had a great time swinging at it blindfolded with an oar from our dinghy. We ended up with quite an audience. I think that the entire visit with Bill, Alissa, and their boys is one special Christmas that we shall always remember.

CHAPTER 44

IN EARLY JANUARY, WE DECIDED to return to Mazatlan. On this trip we decided to make the run without stopping. We weren't going to deal with Isla Isabella again. We left the Puerto Vallarta Marina about twenty hours before a planned early- morning arrival in Mazatlan. As we headed north, in the distance we could see the white guano-covered rock at Isla Isabella in the moonlight. As dawn approached, we were already observing the very bright light in the lighthouse sitting on top of a high prominence at the main harbor entrance at Mazatlan. We were probably fifteen to twenty miles south of the light when we first observed it.

In the El Cid Marina, which is at the north side of Mazatlan, we were again fortunate to be able to get a suitable slip rented for a reasonable rate. On this second Mazatlan stay, we knew very well our way around the area. We decided to expand our horizons by renting a car to explore more of Mexico. The route we chose to follow went east and south. We considered going to Mexico City but the rumors of traffic and the potential of crime caused us to go near, but not to actually enter the city area. Our nearest approach was Morelia, where we toured that city and spent the night.

The next day we used the Autopista (a good toll road) that runs from Mexico City to Acapulco. In Acapulco, we found a beach-front hotel, where we spent the next two or three days. We were generally disappointed in Acapulco as the area around where were staying was older, somewhat run down, and dirty. We were warned about the potential of crime so we avoided venturing out at night.

When I went to fuel the car, the gas station attendant tried to cheat me by trying to charge me for more gasoline than I received. As I argued with the attendant, his buddies came and started to surround me. I quickly got into the car and drove away. It was a very uncomfortable situation to say the least.

The trip back to Mazatlan was a long, and at times difficult, drive. At the time, there were many sections of the road that had only two lanes. There was a lot of slow-moving truck traffic that made it necessary to do a lot of passing. It was difficult to hold an average speed of at least forty miles an hour.

Approaching Puerto Vallarta, we decided to spend the night there before returning to Mazatlan. The little Ford Escort that we had rented had, at the start of our trip, only three thousand kilometers on it and was quite new. When we returned it, it had a lot of squeaks and rattles, because of the pounding it had received on the Mexican roads, which were in many cases not in good repair.

Later in our stay, we were able to enjoy the Mazatlan Marti Gras. Supposedly it is the third largest Marti Gras celebration in the world, behind Rio de Janeiro and New Orleans. Alfred and Elizabeth had friends staying with them. The six of us took a taxi to the area south of town where the celebration was to be held. Before the fireworks and other displays took place, we walked among portable stages and trailers that were set up to accommodate the bands and the food vendors.

Around 10:30 p.m. the fireworks began. Part of the show involved a duel with some large, powerful, lasers. The duel represented the battle that had occurred in the 1800s when a fleet of French warships attempted to capture Mazatlan. The fixed guns in Mazatlan prevailed. The French were never able to succeed in occupying central Mexico.

When our month in Mazatlan was up, we journeyed northwesterly to La Paz. This trip took us across the mouth of the Sea of Cortez and somewhat north in order to enter the harbor area of La Paz. Amazingly, this run, while being followed by Alfred, Elizabeth, and their friends, was so flat the water was simply glassy smooth. Many large turtles were floating on the surface enjoying a siesta in the warm sun and smooth water. Often the turtles had a good-sized gull sitting on top of their exposed shells.

After we had crossed the mouth of the Sea of Cortez, we ran into a dense fog bank as we approached the southern end of the Baja Peninsula. With our radar and G.P.S., we knew our location and could monitor any nearby vessel traffic. Dori had been at the helm, since I was napping in our stateroom. Dori was surprised when she got a radio call from the retired Naval Commander who was, at the time, running Alfred's boat, suggesting that we ought to stop because the fog was limiting our visability. (Dori suspected that the Commander did not want to trust a woman to navigate in the fog.)

By this time, I came on the bridge and realized that Dori had been doing a fine job of piloting. We called Alfred's boat and told him that we were going to continue on our way. Eventually, the fog broke and we made our way to Marina Palmyra, where we rented slips for another month. At that time Marina Palmyra was the best and most modern of several major marinas in La Paz.

The only disadvantage with Marina Palmyra was that it was located several miles north of town. There was a marina shuttle, but it ran infrequently. Coming back from town loaded with bags of groceries was not fun.

This marina was good from the stand point of on shore rest-room facilities. As a result, our holding tank capacity was never an issue. Another nice feature of the marina was that a farmer would come by with his vegetable truck several times a week. The vegetables that he sold were fresh and quite inexpensive. Also, an English language newspaper was delivered to our boat each day. Spanish lessons were available for the marina tenants. Sometimes Dori and Elizabeth attended these classes.

Overall, La Paz was O.K. but not as good as either Mazatlan or Puerto Vallarta had been. While in La Paz, I pushed my running more vigorously. The road to the north of the marina was up hill. The daily runs up this hill got me in good shape.

Toward the end of our stay in La Paz, I took a business trip to France. My meetings were in the south, near Bordeaux. Two Navy men who were attendees at the meeting (one a Navy Captain) were also runners. Before sunrise, the three of us would run through the streets of the city. Although each of them was a good bit younger than I, I had been training so hard that I was able to run them ragged as we covered our four to five mile course.

On this trip Dori was able to accompany me, leaving the boat in the marina at La Paz. On the afternoon of the last day, we drove our rental car from Bordeaux to Paris. Staying overnight at the Paris Airport, the following morning we caught a non-stop flight from Paris to Los Angeles. Arriving in Los Angeles, we then took a flight to the airport located between Cabo San Lucas and La Paz. A rental car was used to get us back to La Paz.

When we left for our trip to France, we had prepared the boat for an immediate departure from La Paz to start our trip home. Tired and jet-lagged, we fired up and started our run around the southern end of Baja. With full fuel tanks, we had enough fuel to make the run from La Paz to Turtle Bay without stopping at Cabo. We were again on an overnighter as we made our way to and around the end of Baja.

Around 4:00 a.m. we were passing by Cabo. The seas became very rough. The waves seemed to be coming from all directions. The tops of the waves I would have guessed to have been ten to twelve feet. These waves were crested with white water, driven by strong winds. I went below to inform Dori, who was sleeping, that we couldn't go on. We needed to turn back to Cabo to wait for better sea conditions.

Turning the boat around was frightening. For a few moments, the boat was caught in a severe beam-sea condition. Our dinghy, stowed on the upper deck, even though securely tied down, began to slide back and forth, making shrieking noises as it did. I feared that it might slide off and that we might lose it. I even felt that the boat might roll. In a moment, I completed my turn and our near panic situation was over. It was unbelievably calm as we made our pre-dawn entry into the Cabo harbor.

Spending most of the day resting, in the late afternoon, we checked weather reports. It appeared by late the next morning the sea conditions should calm down. In the morning, on foot, we proceeded to get all of our departure paper work done. Around 11:00 a.m., we tried again to make our way north. As predicted, sea conditions, while not perfect, were reasonable.

We ran continuously from Cabo to Turtle Bay, a distance of around four hundred miles. In Turtle Bay, we knew the ropes this time and were much less concerned about what we needed to do to refuel. The dock there is on high pilings. Tying up against them could be damaging. Therefore, the procedure is to accomplish what is called a Med tie. In this case, you drop your anchor an appropriate distance from the dock and back in. When you are close enough, you throw two stern lines to people up on the dock. If done properly, your boat rides securely and safely away from the dock. The fueling hose is passed down, and fueling can begin. This time we created some good will by having candy bars to throw up to the kids on the dock. There was no spilled money this time.

Although it was already late in the afternoon, we decided to move on. When we got to Cedros Island, we recalled the poor anchoring at the north end. We decided to anchor at the southern end near a salt mining and shipping operation. By this time, it was dark and with some trepidation, we pulled in close to the salt operation and dropped the hook.

In the morning we got underway, planning to make no further stops. On this leg our engines ran continuously for fifty-two and a half hours. North of Cedros Island we ran into the rough water that is traditionally there. The boat banged and slammed while making only seven to eight knots. The boat took it well, but it was really hard on this already exhausted crew. Even using the bathroom was a chore, as you had to crawl on hands and knees from the flybridge to the head to avoid being thrown and injured.

We heard later that an eighty-foot yacht that had formerly belonged to Julie Andrews had traveled ahead of us on the day before. The rough seas had blown out a port hole on that boat, causing interior water damage.

The only mishap on that portion of the journey was that on the morning that we were passing Ensenada both Vacu-flush toilets stopped working. We were left with only a bucket to use for the next several hours before arriving in San Diego. In San Diego, the toilet system was repaired. It was done in such a way that each toilet was wired independently so that it is now unlikely that both would fail at the same time again.

At this point, we were anxious to get home. We ran north from San Diego using essentially the same stops that we had made traveling south. One difference was that we spent a night anchored in the Coho anchorage, which is just south of Point Conception. Early morning trips around Point Conception usually provide smooth sea conditions, without encountering the notoriously rough water at that dreaded point.

Further north, out of Monterey, we thought that we were home free. As we were passing the coast along Pacifica, looking ahead we saw something on the horizon that we didn't understand. As we got closer, we realized that we were seeing some large waves that were piling up on some shallow waters known as the Potato Patch. Those waves looked ominous. We chose to avoid them by turning out toward the open sea. We hoped that a San Francisco Bay approach in the deep water channel heading toward the Golden Gate would be safer.

Around this time, a Coast Guard helicopter made a low pass over us. We have often wondered what his intentions might have been. Farther out, as we intersected the deep water channel, the waves were still large. Nevertheless, we made our turn and started in. As each wave came in from behind, it looked as if it might swamp our rear cockpit. I had Dori call to me as each wave approached so that I could speed up to stay ahead. I couldn't get too fast because if I topped the crest of the wave I was riding, the boat could plane down the front side with the danger of broaching.

After a tense half hour, we came under the Golden Gate and entered the bay where the conditions were calm. A trip up the bay and rivers in bright sunshine took us home after seven months and four thousand miles of cruising.

CHAPTER 45

AFTER RETURNING HOME, WE REALIZED that there was some Mexico cruising that we had missed. We had failed to proceed farther south than Puerto Vallarta. We should have rounded Cabo Corrientes and explored the picturesque anchorages that were south of the cape, but north of Manzanillo. Also, when we were in La Paz, we failed to proceed northward up into the Sea of Cortez. Accordingly, we began planning another long cruise to the south. The boat had survived the first trip so well that no significant upgrades, maintenance, or repair items needed to be accomplished.

In the fall of 1997, we embarked on the second trip. Our friends Marge and Howard with the Krogen were ready to join us for their second try. As we traveled southward along the California coast, we noticed that the sea conditions were somewhat rougher than they had been in 1995. As we spent several weeks in a marina in San Diego, there was a lot of talk by south-bound cruisers that, because it was an El Nino year, cruising conditions to Mexico could be less favorable. After further consideration, in view of the poor conditions that we might encounter, we decided to abort the Mexico leg and start back up toward home.

Our friends went straight up the coast, but, since our daughter, Linda, and son-in-law, Wayne, were living in Long Beach, we decided to stop in the Long Beach area and visit with them for several weeks. We were fortunate that we were able to rent a very nice slip in Huntington Beach for a month. It was handy to be able to go and visit with our daughter and son-in-law. They had an extra old Toyota, which they were able to loan us, providing us with transportation to roam around the area while we were there.

While we were there, Dori and I were able to help with the businesses that Wayne and Linda were operating at the Long Beach Air Port. They were involved in importing and selling French military jet trainers that had been declared surplus by the French Air Force. My help involved doing some mechanical work on the jets as engines, batteries, and other accessories were removed and replaced. Dori was able to help our daughter work in an office located in a balcony overlooking the large hangar where the parts and jets were stored. Her job involved looking up time remaining before overhaul of various aircraft components listed in the log books and in other documents.

The aircraft business, in addition to importing and selling the jets, was also involved with towing banners and billboards behind a couple of single-engine piston airplanes that they had. The banners that were towed behind these aircraft were sometimes quite large, measuring approximately sixty by thirty feet.

The procedure used for getting these banners into the air involved laying the banner out on the grass between parallel runways and attaching the tow line to a line draped between two poles about ten or twelve feet high. The aircraft would pick up the banner by diving down and having a hook suspended from the tail of the airplane snag the line hanging between the two poles. After the banner had been towed for a while over beaches and sometimes over football stadiums, the plane would return to the airport and drop the banner on the grass by releasing the hook with a control in the cockpit.

This entire process took quite a bit of time and effort on the ground. The banner had to be laid out and folded in a very particular way so that as the aircraft took hold the banner would be lifted into the air incrementally. The work was hot and dusty, making additional hands a welcome addition to the process.

When it came time to leave, we timed our departure so that we would arrive just south of Point Conception in the late afternoon. Because the seas are often very rough rounding Point Conception, our strategy for this part of the journey was to anchor for the night in the Coho Anchorage. Very early in the morning, about 4:00 – 5:00 a.m., we would pull the anchor and proceed around the point, while the winds and seas were calm. This part of our northern trip worked as planned.

As we kept working our way north, the seas roughened. We, therefore, decided to pull into a sheltered but shallow cove at San Simeon. In the cove were several mooring buoys, one of which we tied up to for the night. When we awoke in the morning, we listened to the weather forecast on the radio, and heard that the swells outside were in excess of twenty feet. It was an easy decision to decide to wait another day. Fortunately, the rest of our trip home came off without a hitch.

CHAPTER 46

A FRIEND, WHO ALSO WORKED at the Lab, had acquired a property on the island of Roatan, off the coast of Honduras. This friend had done a considerable amount of mechanical work on a Jeep van that he wanted to put in place at his Honduran property. Our friend who owned the property was still working and was unable to find the time to drive the Jeep down to Roatan. Marge and Howard, our friends with the Krogen, and Dori and I had the time, since we were all retired. We all tend to be rather adventurous so we volunteered to drive the Jeep down through Mexico and part of Central America to Roatan.

In addition to our own luggage, the Jeep was loaded with extra wheels and tires and lots of tools and other equipment. By the time that we got it fully packed, you could not see out the back window, since everything that we were carrying was piled all the way to the ceiling. After we crossed the border into Mexico, we followed Route 200, also known as the Pan American Highway.

About thirty miles below the border into Mexico we encountered a stop for immigration and passport control. Although we had a letter from the owner granting us permission to transport his vehicle and a copy of the certificate of ownership, Mexican customs required that the certificate of ownership be the original. We were turned back and had to spend the night on the U. S. side of the border. Our friend, the owner of the Jeep, was able to drive the document to the San Francisco Airport, where it was given to a Southwest Airlines pilot who got it to us the next day in Tucson. With proper documentation finally in hand, we proceeded onward.

Since we had no time schedule to follow, we sometimes stayed in a place for several days. These stops were a welcome relief because the trip turned out to be very long. The total mileage that we needed to cover was about four thousand miles. As we traveled south of the major coastal cities in Mexico, the condition of the road began to deteriorate.

At one point, we found the road blocked and had to return thirty or forty miles. In another case, we discovered that heavy rain had taken out a bridge that we needed to cross, forcing us to ford a stream that was running through the area below the washed out bridge. Fortunately, the Jeep had large all-terrain tires and four-wheel drive so we made our way through without difficulty.

At the very southern edge of Mexico, we encountered the border crossing into Guatemala. At the border we came across a long line of cars waiting to clear border inspections. We wanted to somehow expedite our way through. We tipped a young boy, who directed us through the heavy traffic, getting us to the inspection station much sooner. The inspections and the paper-work needed were expensive. It was even required that the vehicle be fumigated in order to be allowed to enter Guatemala.

By the time we arrived in Guatemala City, we were all very tired. We, therefore, allowed ourselves to break the budget by taking expensive but luxurious rooms at a high-rise Sheraton Hotel. As we settled into our room on the twenty-second floor, a powerful earthquake shook the building. Dori, was in the shower at the time. At first she wasn't sure whether it was an earthquake or whether she was wobbly from exhaustion. The rolling had been quite strong, but the building was apparently undamaged.

The route we chose when leaving Guatemala and entering

Honduras was a deviation from our main route. We chose the routing in order to visit the ancient ruins at Copan. As we approached the Honduran border, we were prepared for the worst; however, since we were on a back road, there was no traffic and we proceeded through the crossing without delay. The visit to the ruins was outstanding. They were far more impressive than some of the Mayan ruins we had seen in Mexico.

Our next destination was the Honduran coastal city of La Cieba. In La Cieba, we needed to find the means of transporting the Jeep across the open ocean water to the Island of Roatan. After some searching around, we ran into a stranger who said he knew the captain of a boat who would take the van across to Roatan. With a great deal of unease, we gave him the keys to the Jeep and left for the airport to catch a flight from La Cieba to Roatan.

Once in Roatan, we rented a car and drove to our friend's condominium. The next morning we drove from dock to dock looking for the boat that was supposed to have carried the van to Roatan. We didn't know the name of the boat, nor the captain. All we knew was that we thought the boat was painted yellow and green.

Sure enough, after a bit of searching, we found a yellow and green boat. As we looked over the deck, the Jeep was nowhere to be seen. In a short while, some trucks arrived and began carrying away crates and fish traps, clearing portions of the deck. Eventually, enough of the fish traps were removed that we could see the Jeep buried in behind them. We paid the captain for the transport fee and, greatly relieved, we drove the Jeep off the boat and onto the dock. Thus, the four thousand mile transport of the Jeep from northern California to Roatan Island had been completed.

CHAPTER 47

DURING OUR STAY IN MEXICO during the winter of 1995 and 1996, we had made friends with Alfred and Elizabeth, who owned a forty-eight foot motor yacht similar to our own. After their return from Mexico, the couple decided to purchase a larger yacht with the idea of operating a charter business out of the harbor in Mazatlan. Alfred and Elizabeth had become born again Christians while they were in Mazatlan and felt that Mexicans who attended the same church could prove helpful in operating this charter business.

After about a year of operating the business, our friend found that his Mexican employees were taking advantage of him, and he became very disillusioned. As a result, he closed down the business, emptied the house they had rented, and planned to bring the boat back to the U. S. In the meantime, Elizabeth had driven their vehicle and two large dogs back to the U. S.

Alfred sought help for bringing the 64' heavy-duty trawler back north. I volunteered to be one of four of us that would make up the crew. I flew to Mazatlan about a week before our planned departure date. I helped Alfred as he tried to negotiate getting repairs completed that were needed before departure.

As we got closer to our departure date, one of our crew, a retired Navy Lieutenant Commander, cited other responsibilities and thus would not be available to help bring the boat back. The third crew member had very little boating experience, but we thought that he could at least take a turn at the helm while the owner and I were able to get some rest. On the day before we were scheduled to leave, the third crew member got cold feet and decided to back out. That left just the two of us to make the more than one thousand mile journey back to San Diego.

Since Alfred and I had made the trip to the north before, we thought that, although the trip would be rigorous, we should be able to accomplish it successfully. On the day that we departed, we were aware that a hurricane was working its way north in the waters of southern Mexico. As we departed from Mazatlan, making our way across the mouth of the Sea of Cortez, it appeared that some tailings of the hurricane were already appearing as we saw dark clouds and lightening to the north of us. Another disturbing factor was that Alfred, who had a chest cold before our departure, seemed to be getting worse.

We rounded the southern tip of Baja California and began our northern leg. The sea state was at that time O.K. But, the Alfred's physical condition deteriorated some more. He began to spend more and more time in his berth in the master stateroom. This left me piloting the boat more and more on my own. The boat carried more than three thousand gallons of fuel so that a fuel stop anywhere along the Baja coast was unnecessary. We planned to travel twenty-four hours a day.

As we traveled a bit north of Cedros Island, sea conditions worsened. At nine knots, the boat was pounding terribly and I was forced to slow to about seven knots. The pounding had been so severe that radios started coming out of their mountings and were hanging on wires connected to them. In the galley, the refrigerator was working its way out of its mountings and had come about half way out of its cabinetry.

Many of the belongings from the house (such as dishes, pots and pans, and other small utensils) had been packed in cardboard boxes which had been stowed in the forward cabin. Water coming across the forward deck leaked through some of the forward hatches soaking the boxes, which then allowed the contents to fall haphazardly all across the floor.

Around 2:00 a.m., I heard a terrible mechanical–sounding noise. I thought that something might have failed in the drive train of one of the engines. I immediately shut down the engines and pulled the transmissions into neutral. Alfred came out of his stateroom and we both wondered what had happened. Further investigation revealed that the anchor on the foredeck had deployed, taking with it six hundred feet of chain from the anchor locker.

Alfred wanted me to go out on the foredeck to try to retrieve the chain and the anchor. I went out on the deck for a quick look but realized that there was no way that I could operate out there since we were in twelve to fifteen foot tall seas. Had I fallen overboard in those conditions, I would have drowned, since there was no means of rescuing me.

Returning to the pilot house, I told the owner that the only way we could continue was to cut the anchor and chain loose. I went to the anchor locker to investigate and found that the end of the six hundred foot chain was bolted into a wooden beam. With a heavy hammer and chisel, I weakened the beam enough that the chain and the anchor, weighing nearly two thousand pounds, tore loose making a sound like a gun shot.

After many more hours of miserable running in rough seas, we finally began to approach the harbor in San Diego. It was about 9:00 p.m. and I was afraid that the U. S. Customs dock might have been closed. It is against the law to proceed directly to commercial docking before clearing Customs. By this time, the Alfred had become quite ill and needed medical attention as soon as possible. I called the U. S. officials on the radio and explained my situation. The officials asked if I was declaring an emergency, in which case, they would have dispatched a rescue helicopter to take the owner off. I told them that I did not think that was necessary but that I wanted to request a Customs inspector to clear us through as expeditiously as possible.

We tied up to the Customs' dock and very shortly a lady inspector appeared and came on board. She made a rather cursory inspection. She could see the rather bad shape everything was in. We then moved the boat to an empty slip in the harbor and got Alfred the medical attention that he needed.

A year or so later, Alfred died from lung cancer. We have often wondered if we were seeing the beginnings of that disease as we undertook that trip. We later found out that the boat that we had brought north wound up in commercial fishing operations in Juneau, Alaska. Having experienced two difficult boating trips northward from Mexico to California, we had the feeling that we would not want to experience such trips again.

CHAPTER 48

WE HAD BECOME QUITE COMFORTABLE living for an extended period in Mexico and thought that owning a condominium there would be a good idea. We were friends with two other couples who had also indicated an interest in the idea of owning property in Mexico. In order to explore this possibility, we traveled by car from Mazatlan in the north to as far south as Hualtuco. After considering the many possibilities, we found that we liked Puerto Vallarta the best. In Puerto Vallarta, we looked at condominiums in several locations but settled on one located right on the beach, just north of the Marina District.

It was in late 1999 that we closed on the deal to purchase a two bedroom unit, located on the eighth floor of a condominium complex. After we took possession, we engaged an architect (actually a small contractor) to upgrade the unit from the basic condominium that the developer provided. This procedure was common practice in Mexico where the developer built out all of his units as a standard item, and then the owner would make changes as he or she desired.

It took us a while to understand the work habits of the typical Mexican workers. You might think that you had an agreement with them to start on a certain aspect of a project early the following day. Wanting to not miss them, it was not uncommon to sit waiting for them into the late afternoon before they arrived. Their workmanship in stone, like marble or granite, was very good. On the other hand plumbing and wiring often required my supervision to make sure that the job was accomplished to my satisfaction.

In one instance, some plumbing work created an amusing situation. I had contracted for the installation of a water filtration and sterilization system. I carefully monitored the work that the installer had done. Although the building had been completely plumbed with copper piping, the filter system was installed with plastic pipe. During this operation, all the water to our condominium had to be turned off. When the job was finished, the installer advised that the water pressure to our condominium should not be turned on for several hours, since the adhesive bonding the plastic pipe needed to harden.

I agreed with his recommendation. After several hours, I turned our condo water pressure back on as had been recommended. I had gone back into the condo and had been sitting quietly reading alone since Dori had been out entertaining the wives of some of our friends. Suddenly, I heard lots of noise and lots of shouting. I went to the outside hallway and saw a number of workers running up and down the halls. There was a great amount of water cascading down from the floors above.

At about the same time, I observed Dori and her friends, who had just gotten off the elevator, coming down the hallway in the midst of the chaos. I opened the door to the closet where the water filter system had been installed. What, of course, had happened was that one of the glue joints on the plastic-pipe fittings had failed, allowing high pressure water to spew up the plumbing shaft, several stories above where our condo was located.

I later learned that water pressure in the building was quite high, and that the use of plastic pipe was not advisable. When I had another installation done a couple of years later, I made sure that everything was plumbed in copper.

In the 1999 time frame, the availability of quality furniture in

Puerto Vallarta was limited. We elected to do our major shopping for furniture in Guadalajara. As the crow flies, Guadalajara is only one hundred and ten miles from Puerto Vallarta. However, because the mountains of the Sierra Madre lie between the two cities, the drive is indirect and somewhat hazardous in that lots of it is on a windy two-lane road. The car trip normally took over three hours.

In Guadalajara, we were joined by our friends Bruce and Dawn from Discovery Bay, who had also purchased a condo in the same building. We elected to stay in an old hotel lying just off one of the main plazas in the downtown area. As we shopped for furniture, we realized that it was important to avoid anything containing softer woods, like pine, which were vulnerable to termites. Thus, furniture needed to be either made from hard woods or from other material such as rattan.

As we searched, we came upon a large show room with a factory behind where high-quality rattan furniture was on display. The light-colored rattan that we chose had a modern tropical appearance which went well with the environment we would be living in.

Although many people think that Mexican furniture ought to be cheap, it was in fact relatively expensive. As we added up all the prices of the individual pieces that we ordered, we were surprised at how large a bill we had run up. Delivery schedules slipped, but by the time everything was delivered, the furniture fit nicely the décor that we had selected for the condo.

With our furnished condo completed, it was our plan to use it about five months or so a year. We planned to occupy it from early November to about the end of March. Typically we would go home for two breaks. One during the Christmas holidays, and the other in late January or early February in order to catch up on things at home and to take care of mail and other paperwork.

CHAPTER 49

IN 1998, I DECIDED TO again pursue my hobby of flying. I rejoined the Flying Particles flying club, the same club that I had joined before in 1961 when I obtained my pilot's license. I had become an inactive pilot during the mid 1970s because I was pursuing my boating interests, but also because of the funding demands of preparing to put our three children through college.

In order to reactivate my flying status, I had to complete a medical examination and I needed a thorough check-out by a certified flight instructor. Although I had not flown for twenty-three years, my instructor signed me off after only two hours of recurrent training. I flew with the Flying Particles for a short while, improving my flying skills and taking short trips around the bay area.

Somewhat later, Wayne and Linda had taken a small two-place single engine airplane in trade as partial payment for an airplane they had sold to an individual. Since we had loaned their business some money, they gave us the little airplane as a partial pay-back of the loan. The aircraft was a Beechcraft Skipper. We kept the Skipper in a hangar we rented at the Byron Airport. The Skipper was in nice condition and it served us well. We used it for taking some short trips, but we mainly used it to fly to San Luis Obispo and Santa Rosa to visit our daughters and their families.

Our son who lived in Lafayette, California, also became interested in flying. He started taking lessons at the airport in Concord, California. Because we planned to spend the winter months in Mexico, we let our son Bill use the Skipper to complete his training and get his license. The Skipper was actually designed to be a trainer so its use by our son turned out well.

By 2001, the Skipper's performance in terms of speed, range, and passenger capacity were limiting its use. I began shopping for a four-place airplane with more power, speed, and range. The Skipper was put up for sale. We were contacted by a number of potential buyers, but none were sufficiently interested to make a purchase.

Finally, we were contacted by a pair of student pilots from the Atlanta, Georgia, area. Since they were students, they were unable to come to California to take delivery of the aircraft. Dori, who happened to be the one taking their call, volunteered that, if they would buy it, we would deliver it to Atlanta. The buyers had an uncle who lived in California who was an aircraft mechanic. The students had the uncle come to Byron to inspect the Skipper. Because of his approval, and due to our willingness to fly the Skipper from California to Georgia, the buyers felt assured that the plane was O.K.

In early June 2001, we started our journey east. Although neither Dori nor I are particularly heavy, with our weight and full fuel we could only carry sixteen pounds of luggage in order to stay within the legal weight limit. With my flight briefcase and charts, our small overnight flight bag put us a few pounds over gross. In the west, where the terrain tended to be higher, and with the high June temperatures, our take off performance was of concern.

Our departure after a fuel stop in Twenty-nine Palms, California, was a bit exciting. Our rate of climb out of the airport barely stayed above the slightly rising terrain at the western end of the field. We were relieved as we traveled farther east where density altitude figures were much closer to sea level.

Our last hop from Anniston, Georgia, to Peach Tree Falcon Airport south of Atlanta was marginal VFR. We stayed low to remain below the overcast and made the final fifty to one hundred miles in light rain. We were a bit sad to see our little Skipper go, but we were relieved that we had made a successful delivery.

CHAPTER 50

BEFORE I LEFT CALIFORNIA, I had located a Piper Cherokee 235 that was for sale in Healdsburg, California. Although it was a 1973 model, it had relatively low flight time (1,750 hours) and was in excellent condition. After a pre-buy inspection by a mechanic friend from Byron, we bought the plane. The first time that we took off from Healdsburg with our new Piper I felt that we had made the right decision. Its rate of climb compared with that of the Skipper seemed phenomenal to us. We realized that we had bought a great-performing long-range airplane.

Our first long-range trip with it was from Byron to Puerto Vallarta. We flew it down just after the Christmas holidays in 2001. Our border crossing into Mexico at Mexicali was uneventful. However, the paperwork and obtaining permits took a lot of time. Our first fuel stop in Mexico was planned for Guaymas. Because of the time of year, the days were short. When we were still about one hundred miles from Guaymas, the sun went down. In Mexico, it is not legal to fly after sundown unless you are on an IFR flight plan.

The Mexican flight controller called on the radio warning me about my tardiness in making our landing in Guaymas. I advised him that the Mexican officials in Mexicali had delayed our departure. By the time we arrived in Guaymas air space, it was completely dark.

The airport had the typical green and white rotating beacon that we were able to home in on. There were no lights on the field, except a single red light marking one end of the runway. As we got low, tracking the single red light, my landing lights picked up the runway. We made a good safe landing. I felt uneasy parking the airplane in the dark next to other airplanes which had propeller locks on them. It seemed that theft could be a problem.

Nevertheless, we tied up and went to the terminal building where we found Mexican officials waiting for us so that they could examine our paperwork and collect fees. Due to our late arrival, we paid extra over-time fees for the officials who had waited there for us. We collected our luggage and took a cab to a nice clean motel.

The next morning, we filled the fuel tanks and planned to make the remainder of the trip non-stop to Puerto Vallarta. At times, our direct route took us a bit over the ocean, but I stayed close enough to the shore line so that, if the engine failed, I would be able to glide to the beach.

On the way down we encountered moderate southwest winds which slowed our progress and caused us to burn a bit more fuel than we had planned. When we arrived in Puerto Vallarta air space, we had only fourteen gallons of fuel remaining. With this amount of fuel, we would have had about an hour of flight time available to seek another landing site if necessary.

It was afternoon as we approached Puerto Vallarta Airport. Some low clouds were already beginning to move in from the ocean. We had to remain at an altitude of 1,200 feet in order to stay under the cloud layer. Puerto Vallarta is a rather busy airport. Late in the afternoon is when many of the commercial flights are arriving from northern cities in the U. S.

As we started our down-wind entry into the pattern, we could see a Boeing 757 just touching down. At that point, the tower was clearing another jet on final for his landing. The tower asked me to extend my down-wind leg about six miles. The controller in the tower said that he would call my turn to base. When the controller made that call, I knew that he had another jet on about a ten-mile final. I knew that by the time I landed the incoming jet would be close behind.

As I got to about a one mile final, the tower asked if I could land long (the runway is 10,000 feet long). I advised that I could, and I planned to make my touch down 5,000 – 6,000 feet toward the end of the runway. It seemed as if the controller must have been concerned about the spacing, with the jet over-taking from behind. A moment later the controller called and asked if I could land short. Fortunately, my Piper has lots of flaps that are part of a STOL Kit that had been installed on the airplane. I advised the tower controller that I would land short and quickly dumped the excess altitude I had been carrying for the planned long landing.

Our short landing was successful. As we made the first turn-off from the active runway and looked back to the approach-end of the runway, we observed the white smoke coming from the wheels of a 737 that had just touched down. I don't think that a tower controller in the U. S. would have allowed such tight spacing between two aircraft.

When we taxied to the parking area and hangars where private aircraft are kept, we were met by custom officials and friendly-helpful people who worked in the facility where our plane was to be parked. The daily parking fee was high and, if one had wanted to stay for two or three weeks, total daily fees would have been quite expensive. Fortunately, I was able to negotiate a monthly rate that was far better.

While the plane was in Mexico, we took a trip to Guadalajara. We were accompanied by two lady friends who also had condominiums in the same complex where our condo was. One of the ladies had been a very experienced pilot, having been a Beechcraft sales person for a large dealer in Albuquerque, New Mexico. She was type-rated in the twin-turbine Beechcraft King Air. Because of her interest and experience, she flew the right seat.

The trip went fine and the climb capabililty of our Piper was put to good use in clearing the 8,000 foot ridge of Sierra Madre Mountains to the east of Puerto Vallarta. After landing in Guadalajara, we rented a car and drove to a quaint bed and breakfast in Tulaquepaque. After some local shopping and enjoying some good Mexican food, we returned to Puerto Vallarta the next day.

Since the Puerto Vallarta Airport is so close to the beach, I was concerned about the corrosive effect of the salt mist coming from the ocean. After about a month in Puerto Vallarta, I decided to take the airplane north so that I could put it back in its hangar in Byron. Once again we planned to make an overnight stop in Guaymas. Since our fueling situation from Guaymas to Puerto Vallarta had been a little tighter than I had wanted, we made an intermediate fuel stop on our trip north.

As we approached Guaymas, winds were reported to be blowing at about thirty knots. I was concerned because, if there was a substantial cross-wind component, it would have been near the limits of what my aircraft could handle. Fortunately, as we turned on to our final approach, the winds were only about ten to twenty degrees off the runway heading. The landing was uneventful, and after securely tying the airplane down, we headed to the same motel that we had stayed at before.

Strong winds were still predicted for northern Mexico and southern California the next day. When we took off the next morning, however, the winds at Guaymas were practically calm. The winds were also not a factor after we crossed the U. S. -Mexico border landing at Calexico. The border crossing went by without a hitch, as I had made our preliminary call to U. S. Immigration one hour in advance of landing, as required. Upon landing, the friendly U. S. agent came out to the plane and asked whether we were carrying over $10,000 or any drugs. We replied in the negative. He took us inside to sign some papers and never even looked at our passports. After fueling, we were quickly on our way back home.

CHAPTER 51

ALTHOUGH OUR FLYING EXPERIENCE TO Mexico went well, I was not happy about having to keep our airplane outside in the corrosive salt air environment at Puerto Vallarta. During the time that our plane was in Mexico, we had befriended a Mexican couple Joel and Lettie who were running an air-taxi business out of the Puerto Vallarta Airport. Their operation consisted of a number of different types of aircraft. The planes ranged from an old Cessna 150 to a twin-engine Cessna 421. In the absence of my airplane, Joel suggested that I could rent his airplanes as necessary.

One of our first encounters with their aircraft was in connection with an air show and display in the town of Tepic about one hundred miles north of Puerto Vallarta. Planes were to be coming from all around the area. Our Mexican friends wanted Puerto Vallarta to be well-represented at the show. Dori and I were invited to use their Cessna 150. The owners planned to fly the Cessna Cardinal, while another friend would fly the Cessna 206.

As we approached Tepic, the tower was called and permission was granted for the three Puerto Vallarta aircraft to buzz the tower before landing. I found it hard to believe that such shenanigans were permitted at Tepic since it is a good-sized commercial airport.

After we landed, we watched a variety of other aircraft coming in, even including several Piper Pawnee crop dusters. All of the parked aircraft were on display. There were also a number of radio-controlled model airplanes, which were flown for the entertainment of the audience. In the afternoon, we heard that pilots could volunteer to take visiting children for short airplane rides.

I volunteered to fly the Cessna 150 on the rides. I was amazed as the Mexican families brought their little children up to the airplane and stuffed them in without even having sufficient number of seatbelts. I even had very small children riding in the baggage compartment, sitting on top of our luggage. Around and around I went for quite a while, taking these wide-eyed little tots on their first flight and perhaps the last that they would ever have in their life time. I often wondered what the parents and these little children must have thought as the cockpit door opened and there sat this older gringo pilot. It was our observation that we seemed to be the only non-Mexicans in attendance that day.

When the flying ended, the participants were offered a free dinner of assorted Mexican food and free beer. There was a large tray of some kind of barbequed meat. We all had some, plus rice and beans. When I went back for a second helping, a skull of some small animal emerged from the center of the tray. We had apparently eaten roasted goat.

During the dinner, while we were all drinking beer, we noticed that the crop duster pilots, who were sitting together at an adjacent table, were augmenting their beer drinking with tequila from a bottle on the table. While there was still some daylight, they jumped into their Piper Pawnees and flew away with an unknown blood alcohol level.

As our party ended, a bunch of us climbed into the back of a pickup truck and were driven to a hotel a good distance away in the center of Tepic. The next morning we departed on our own and flew the 150 back to Puerto Vallarta.

I used that 150 a number of times taking friends from our

condominium complex for sight-seeing rides along the very picturesque coast line north of Puerto Vallarta. As I kept renting the aircraft, the air taxi operator encouraged me to get a Mexican pilots license. With a Mexican license, I would be able to use aircraft with Mexican registry. The 150 that I had been using was on loan from a friend in the U. S. and thus still had U. S. registry. My Mexican friend said that getting the license would be very easy, and that most of the paperwork involved would be processed through the Commandant's office at the Puerto Vallarta Airport.

One of the first steps was to get a medical examination from the examiner's facility in Guadalajara. Our friends, Hal and Pat, were visiting us from Discovery Bay so we decided to drive to Guadalajara, do some sight-seeing, shopping, and get my medical examination done.

When I went to get the examination, I found that, on the initial examine, one had to obtain a brain-wave scan. Such scans are only accomplished at very specialized facilities, which we had to track down. When we finally located a facility, we had a long wait before I was taken. The exam itself was relatively simple, not very different from an EKG, in that electrodes are attached to your scalp, while brain waves are recorded on a paper chart.

The remainder of the physical is more rigorous than a U. S. medical in that it also includes, what the examiners call, a psychological test. The test is normally a written test in Spanish. Because of my lack of Spanish proficiency, I was given a test that involved drawing pictures and figures. Rather than it being a psychological test, it seemed to be more a test of spatial relationships and perceptions.

The rest of the physical was somewhat routine, consisting of a blood test, a urine test, and an EKG. At the end of the examination, when you had your final interview with the senior doctor, a two hundred peso tip tended to assure a favorable outcome on the medical report.

The remainder of the work needed for getting a license entailed getting over a number of bureaucratic hurdles. First, they wanted copies of every page of my pilot's log book. My pages dated back to 1961 when I first got my license. In addition to making copies, they wanted each page to be notarized. When that was accomplished, they wanted the entire package to be apostled. We had never heard of the process called apostling. When we looked into it, we found that the apostling process was done in a specific office in Sacramento. After submitting our documents and paying a nominal fee, we were handed a very impressive embossed paper indicating that our material had been apostled.

All of this material was returned the Puerto Vallarta Commandant's office along with a medical examiners statement and a statement from a flight school in Guadalajara that I had been properly instructed on aircraft safety procedures. When all of the many fees were tallied up, it turned out that my license acquisition cost about $1,000 and lots of frustrating effort to clear all of the hurdles that had been imposed on us. In spite of all this, I kept an active Mexican license for about four years.

CHAPTER 52

ONE DAY WHILE WE WERE sitting beside the very large infinity pool at our condominium complex, the developer of the entire complex came walking by with several other men who had been helping to run the business. As we spoke to him, he said that we really should consider buying a new larger condominium in another tower that was just being constructed. He was offering a pre-construction price that sounded very attractive. With hopes that our family and friends would join us more often, we decided that we should move up from a two- bedroom to a three-bedroom unit. Looking over the availability of units for sale, we elected to buy a three bedroom penthouse on the 13th floor.

As the upper units approached completion, we engaged a contractor in order to make a number of significant changes in our unit. In the laundry room we added cabinetry and a sink. The arrangement of the cabinets in the Kitchen was altered. Marble countertops were installed instead of the standard Formica. The sink and the faucet were also upgraded.

Since we had an end unit, we were able to significantly enlarge the front entry to the condo. We had a beautiful marble-topped bar built with glass shelves behind. We added a built-in entertainment center, which provided space for our TV, electronic components and other storage areas. Its beauty was enhanced by white plaster columns on each side and lighted niches where we placed ornamental glass. The entire wall across the room from the entertainment center was mirrored.

Behind the second bedroom was a circular cupola that in the standard unit was not weather tight. We added glass windows and a built-in desk, with a marble top, so that we could use the space as an office. This revamped cupola was a very pleasant place to work since it had wonderful views of the beach, ocean, and adjacent golf course.

We were fortunate to have sold our two-bedroom unit to a couple we had been friendly with. They had had a one bedroom unit, and like us were upgrading their property. We had only been in our penthouse a short time when a strong earthquake about one hundred miles to the south struck the coastal areas of Mexico.

Our friends, Helga and Mark, were visiting us. As we returned from eating dinner at a restaurant, we boarded the elevator to go up the thirteenth floor. As the elevator was passing the ninth floor, the elevator cab began to shake rather strongly. At first, we thought that it was a problem with the elevator, but then we realized that we were in the midst of a sizable earthquake. As we arrived at the thirteenth floor, the elevator doors only opened a crack. I was able to use my hands to pull the doors sufficiently open that we could all exit the cab.

The shaking continued as we entered the hallway. We encountered a couple from Chicago, coming down the hallway, holding on to the walls in fright. The couple had never experienced an earthquake before and wondered what to do. In a few more minutes, the shaking died down and we continued down the hallway to enter our unit. Upon entering, we were surprised at the amount of damage that our decorative items had sustained. Pieces of pottery and glassware were shattered and scattered all over the floor.

We noticed water coming from each of the three bathrooms and feared that some of our plumbing had broken. What had actually happened was that the water in each of the toilet bowls had been sloshed out by the violence of the building motion. Being on the thirteenth floor had some disadvantages. After we got over the initial shock of feeling and seeing what had happened, the four of us got together and cleaned up the mess on the floor. We estimated that the damage to our decorative items was probably in the order of one thousand dollars and thus was not sufficient to be covered by the insurance we carried.

There was no apparent damage to the basic structure of the condo or to the building. The exception to that was that some sliding joints between the closely-spaced buildings had broken loose, as designed. Repair to those joints was not particularly significant. The elevators remained inoperative for about two days until technicians could be sure that there was no mechanical damage.

A week or so later we rented a Cessna Cardinal from my air taxi friend and flew to Guadalajara to shop for replacement of the glass and pottery items that had been destroyed. As we returned to Puerto Vallarta, we had the baggage compartment and the back seat loaded with items that we had purchased. We had even stuffed a wrought-iron framed mirror about four feet long lengthwise across the back seat.

Our new penthouse was a joy to use for entertaining. We hosted several large cocktail parties, sometimes with thirty or more guests. So that I could spend more time socializing with everybody, we typically hired a bar tender to take care of serving drinks. Our life in Puerto Vallarta tended to be more social than it was when we were at home. Every day there was visiting and chatting with friends and neighbors by the swimming pool.

North of our condo complex the ocean beach area was completely undeveloped. The long stretch of undeveloped beach was great for taking long afternoon walks. In the fall while the ocean water temperature was still warm we sometimes enjoyed swimming in the ocean.

CHAPTER 53

FROM THE VERY BEGINNING OF our stays in the Puerto Vallarta condos, our grandsons, Tim and Chris, would come to visit us for a week when school was out on Thanksgiving break. While they were still quite young, Billy and Alissa would put them on a non-stop flight from San Francisco and Dori and I would pick them up at the airport as soon as they landed. As required, they carried a letter of permission from their parents, which also specified who was to pick them up when they arrived at their destination. A similar process was used when they prepared to return home.

These Thanksgiving visits became regular. To have fun and to keep the boys occupied, we developed a routine of activities that both we and the boys looked forward to. Just outside of town there was a great water-slide park. A couple of the slides were quite high and fast and we all enjoyed riding them. The boys never got tired of pushing us in our floats around the river, trying to steer us under falls.

Another favorite activity was riding Go-Carts around a great Go-Cart track just north of town. PopPop was always the driving champ, but as the boys became older, they became more of a challenge.

There were also places that rented dune buggies and ATVs. Those turned out to be great fun because with them we were able to explore some of the back country well outside of Puerto Vallarta and some of the mountainous regions that were basically foothills of the Sierra Madre Mountains.

With the dune buggies, we climbed trails that were high enough that the air actually became cool. The VW powered dune buggies that we rented were not in the best of shape, making them a bit of a challenge to drive. You couldn't help but get dirty as the line of buggies kicked up dust and sometime mud as we sometimes crossed small streams. Unfortunately, Dori and Tim got more than their fair share of dirt in one episode, as manure was thrown up on their clothes.

On one of the later visits, our son Bill accompanied Tim and Chris to Puerto Vallarta. We had heard people talk about how great rides on the so-called zip lines were. We elected to give them a try, the site that we chose was south of town and had a course consisting of nine sequential runs. Each of the runs was approximately a quarter of a mile in length. In several cases, the runs crossed canyons that were three to five hundred feet deep.

The cable trolleys seemed quite secure and in good repair. As long as you had no fear of the height, the high-speed runs from point to point were fun and exhilarating. Several runs passed over a deep canyon with huge granite boulders and a beautiful stream running beneath. All of these activities provided us with many good times that we enjoyed with our Tim and Chris. We all have fond memories of them.

CHAPTER 54

WE USED OUR PIPER CHEROKEE for some fairly long trips. Carolyn and Dick, who were friends of ours in Discovery Bay, had purchased a forty-five foot Bayliner in the Pacific Northwest. They invited us to join them for about a ten-day cruise from Campbell River, British Columbia.

Campbell River is approximately half-way up Vancouver Island. Thus, it is around two to three hundred miles northwest of Seattle. After leaving Discovery Bay in the Cherokee, we made two fuel stops along the way. One was in Medford, Oregon, and the other was at another airport near Gig Harbor, Washington. After the second stop, we had to navigate carefully through the restricted areas due to the large amount of naval activity and the class B air space surrounding the Sea-Tac Airport. Farther north, we flew into Canadian airspace and flew along the eastern coast of Vancouver Island until we finally arrived at Campbell River.

We encountered no difficulties in clearing Canadian customs and immigration. We were ahead of schedule in our arrival at Campbell River, allowing us several days to tour the area by car. We drove down to Victoria and joined some friends from our Mexican condo complex who lived in the area. In fact, we made an overnight stay at one of the homes located in Arbutus Ridge, just to the north of Victoria. Because the exchange rate at the time was so favorable to the U. S. dollar, we thought about the idea of building a custom home in this beautiful location. Like so many such ideas, we pursued it no further.

Returning to Campbell River, we joined our friends Carolyn and Dick on their Bayliner. The cruising in that area is always outstanding. We had a great trip and revisited many places we had seen many years before.

Preparing to leave for home, we dropped off the rental car at the Campbell River Airport. When we landed and tied the plane down, we had noticed a fair amount of bird activity in the grassy airport parking area. A normal pre-flight inspection of an airplane's engine involves only checking the oil, the security of the propeller, and the tightness of the alternator belt. Checking the oil is done through a small access panel, located on the top of the engine cowling. Since this opening is only about four inches square, only a small area of the engine can be observed.

Because of the bird activity that I had seen before, I made the decision to remove the entire upper half of the cowling over the engine. With the cowling removed, we were shocked to see a large bird nest sitting on top of the engine, blocking air flow to two rear cylinders. The nest was about fifteen inches in diameter and contained several light blue eggs. To Dori's dismay, the mechanic took the nest and threw it in the trash. However, had the nest remained, the engine would have been damaged and there was a potential for fire.

Removing the nest and its associated debris took a fair amount of time since straw and other debris was lodged in the cooling fins of the rear cylinders. The mechanic brought a tank of compressed air and supplied us with tools to help finish the cleaning process. Reinstalling the cowl, we were soon on our way back home.

CHAPTER 55

I CONTINUED TO USE MY Mexican pilot's license in Puerto Vallarta, as we occasionally rented the air-taxi operator's airplanes. These occasional trips made an easy one-hour journey to Guadalajara, rather than the more than three hour dangerous drive on the two-lane road between the cities.

On one trip I few to Guadalajara alone to take the annual medical examination in the facility in Guadalajara. My air-taxi friend Joel had purchased a Cessna 182 RG, which had been repaired after suffering damage from a very rough landing on an unimproved field. Joel wanted me to fly it to Guadalajara to have the mechanics there check out the landing gear lights, which were inoperative. The plane had a small mirror mounted to the wing strut so that you could more or less monitor the retraction and extension of the landing gear, even though the lights were not working.

My thorough pre-flight inspection of the plane seemed to show that the repair of the damage had been reasonably-well performed. I topped off the fuel tanks and flew to Guadalajara. As I approached Guadalajara, I called both approach control and the tower but received no response. I repeated my calls a number of times but heard nothing. I decided to fly directly at the tower, flashing my landing lights, hoping that the tower would recognize my problem and give me landing clearance by flashing a green light. No green light ever appeared. I repeated my flashing landing light procedure several times, but I got no response.

Since I had important reasons to want to land, I waited until there was no visible traffic approaching the runway and proceeded to make a landing. As I taxied to the parking area, an official-looking car pulled up beside my airplane. I was anticipating a reprimand for having landed without permission. As I began to explain my situation to the official, he responding by saying, "It's no problem. You're in Mexico."

After making contact with the repair facility, I taxied the plane about a mile to where the facility was located. I told them about the lack of landing-gear lights, and I explained that my radio transmitter seemed to be inoperative. With some uneasiness, I gave the mechanics the keys to the aircraft. I headed for the medical facility in order to complete my medical examination.

The medical examination process typically takes about half a day, so one is generally finished by lunch time. In this case, however, the computer system broke down, which didn't allow the paperwork that I needed to be completed. It was nearly 3:00 p.m. when I finally had the papers that I needed in hand.

Returning to the hangar where the repairs were to be done, I was told that the small activating lever that caused the landing-light switch to function was broken. It needed a new part that was not available, and would not be available for about a week. They made a check of the radio transmitter and found that it seemed to be working.

With paperwork from the repair facility in hand, I made the mile walk back to the terminal to activate a flight plan so that I could get clearance to fly back to Puerto Vallarta. One of the officials there said that he needed to speak to one of the mechanics before he would sign-off on my flight plan, even though I had a written description of what had been done.

Back to the repair hangar I went, explaining to the mechanics the problem that I had faced. This time, I was driven back to the terminal by the man who had looked at the radio. The official said that this person was inadequate and that he needed to talk to the man who had done the inspection of the landing-light problem.

By this time, I was noticing that dark clouds were appearing over the mountains to the north of the field. Occasional flashes of lightening could also be seen. As I had departed Puerto Vallarta, Joel had said that he needed the airplane back for another flight that was to be made at 5:00 p.m. By this time, with the approaching weather, and the continuing delay of my departure, I was very concerned about being able to make the 5:00 p.m. return schedule.

I made a strong appeal to the official that had been holding up my flight plan approval. He finally relented, allowing me to go. I took off with landing gear lights still not operative, and with a radio transmitter whose operation was highly suspect. As I climbed away from Guadalajara, I attempted to call departure control. Once again, I got no response. The radio did not seem to be working, and I now had to worry about entering Puerto Vallarta air space and traffic without an operating radio.

As I flew toward the west, the dark clouds that had appeared to the north were moving into my flight path. Heavy rain began, accompanied by some mixed-in sleet. I began to alter my course to the south. The cloud deck ahead had lowered to the point where it was not far above the Sierra Madre peaks that I had to clear. As I crossed each ridge, I made sure that I could see daylight out the other side while just staying clear of the clouds above. Because of the approaching weather, I was now about thirty degrees south of the course that I had intended to fly. As I continued on this path, it meant that I would arrive over the western coast line roughly thirty miles south of where I wanted to be.

I finally left the dark clouds and the rain behind and found myself on top of white broken clouds over the coastal area. I flew to the north in preparation of landing at the Puerto Vallarta Airport. As I approached Puerto Vallarta, I called the tower, and to my amazement, they answered, clearing me for the approach and landing.

I taxied to the place where the air-taxi operator kept his airplanes. As I shut down, Joel ran out and told me that he was very glad to see me, since TV news had reported that there had been a fatal airplane crash in the vicinity of where I had just been flying. As it turned out, a small twin-engine airplane, carrying the governor of the state of Colima and several others, had crashed due to weather, killing all aboard. We suspect that that airplane was flying on an IFR flight plan and had flown inadvertently into a cell of severe turbulence in the storm I had recently flown under.

When I went back to our condo, Dori had been unaware of the accident that had occurred, but she was very concerned because I was so much later than the arrival that I should have had. After this rather disturbing episode, I decided that the condition of these Mexican aircraft that I had been flying was potentially dangerous, and that dealing with Mexican officials could at times be unreasonable. Although we had enjoyed some fun times using the Mexican aircraft, we thought that we should abandon any further use of them.

CHAPTER 56

IT MIGHT SEEM THAT BECAUSE of all of the activities that I have been recounting about our time in Mexico, that there wasn't much going on at home or at the Livermore Lab. Our stays in Mexico were generally less than half a year, which left the remaining time to be available for activities at home and at the Lab. At the Lab, I became actively engaged in preparing two historical technical documents.

One covered the history of a particular class of fission-weapon primaries. The design and development of this class of primaries was begun in 1961 and was widely-used many times in full-scale nuclear tests at the Nevada Test Site. Two members of this class of systems entered the U. S. stockpile. One was the primary for the large high-yield warhead developed for the Spartan (ABM) Missile. The other was for the medium-range tactical missile called Lance. The report that I prepared included technical discussions of the developmental paths that were followed as the program proceeded. The report provided figures and data sufficient to give younger designers information of how to proceed should programs of this type need to be resurrected.

Another comprehensive report that I prepared covered the

design, development, and testing of nuclear artillery shells. Such shells began to be developed in the early 1950s. The first was the 280 mm diameter MK9. The MK9 is one of only a few systems that had had a full-scale operational nuclear test.

The MK9 was fired from a cannon at the Nevada Test Site in 1953. The nuclear explosion, which took place seventeen miles down-range, produced a measured yield of approximately 15 kt. As was discussed earlier, nuclear artillery effort continued many years making the projectiles smaller, more useful, and more effective.

In the artillery shell report, I covered not only the nuclear designs worked on at Livermore, but also those which were worked on at Los Alamos. In the case of the Livermore systems, two of them entered the U. S. stockpile. One was a 155 mm projectile named the W48. The other was the 8" W79, which had enhanced radiation capability. As I had mentioned earlier, I had served as program leader for the W79 for a number of years.

Since I was only working part time, the preparation of these two extensive documents took some time to complete. For my work in preparing these documents, as well as other work I had completed throughout my career at the Lab, I was presented with the Weapons Recognition of Excellence Award by the Department of Energy in September of 2002.

In June of 2002, the Livermore Laboratory passed a significant milestone in its weapons and scientific efforts. The Laboratory celebrated its Fiftieth Anniversary from its inception in 1952. As part of the celebration of this anniversary, a series of classified talks was presented to an audience of senior DOE officials, and senior retirees from the Laboratory. I was invited to give one of these classified talks regarding the many projects in which I had been involved.

In addition to writing these comprehensive reports, I periodically put together and delivered briefings about my past work to some of the physics divisions and weapons-engineering groups. I have also been invited to give the briefings to other agencies in Washington, D. C. involved with weapons assessments and weapons technological capabilities. These activities have continued up to the present time.

The weapons program maintains a highly-secure vault where very accurate full-scale cut-away models of all the warheads LLNL has been involved with are kept. Senior people visiting the Lab with the proper clearances and a need-to-know are sometimes given guided tours of this vault by program managers and a number of docents. As a portion of my part-time work, I have in the past and continue to serve as one of these docents. Another portion of my continuing part-time work involves the informal mentoring and answering questions raised by some of the younger B Division primary designers.

CHAPTER 57

FOLLOWING OUR RETIREMENT IN 1991, Dori and I joined the Discovery Bay Athletic Club. We follow a six-day a week regimen where we use the treadmills for twenty to thirty minutes, and then use a combination of machines and free weights to work out our upper and lower body. When we started on the treadmills, I would run on them at about six miles an hour. My near thirty years of running outside on paved roads had taken its toll on my spine. On the treadmills I realized that I could burn calories at the same rate by briskly walking on a steep incline, rather than the pounding my back was subjected to when running on the level.

Since we started this routine back in 1991, we have continued to follow it rigorously. Normally, we are out of bed by 5:30 a.m. We eat a light breakfast, read the morning paper, enjoying a cup of coffee, and then we head to gym at 7:00 a.m. for an hour's morning workout. Although following such a schedule takes serious discipline, we realize that it is one of the most important things that we can do to maintain vigorous health. Our efforts may not necessarily prolong our lives. However, because of the strength and stamina that we are able to maintain, we are able to continue in engaging in all of the activities that we were engaged in when we were younger.

CHAPTER 58

BEYOND THE TRIPS THAT WE have taken flying our Cherokee to Canada and Mexico, we have taken some other significant trips within the U. S. In one trip we flew to Sun River, Oregon, where we had rented a nice cabin that had facilities adequate for accommodating Dori and I and our son Bill, his wife Alissa, and two of our grandsons, Tim and Chris. By this time Bill had earned his private pilot's license and had joined a flying club based at Buchanan Field in Concord. Dori and I arrived a few days early and rented a van large enough to accommodate all of us.

Bill and his family arrived a few days later in a club 182 RG. The cabins at Sun River are rustic and sit amongst groves of large trees. The cabin contained six bikes, enough for everyone. There are lots of paved trails through the woods and fields in the surrounding area where we enjoyed bike riding. There were raft rentals on the Deshutes River. We rented two rafts and spent an enjoyable day drifting several miles down the river, where we were picked up and taken back to the resort.

We had a similar experience where Dori and I flew to Flagstaff, Arizona, and we were joined by Bill and his family a couple of days later. Dori and I broke up our trip to Flagstaff by making an overnight stop in Laughlin, Nevada. In many respects, Laughlin was similar to Las Vegas. There are large hotels with onsite casinos and many restaurants. An advantage of Laughlin is that everything is less expensive compared to Las Vegas.

On the day that Bill arrived in the 182 RG, a thunderstorm was approaching, and they got on the ground just before it began to pour. With the large van that we had rented, we took in the sights at the Grand Canyon and Sedona. The drive between Flagstaff and Sedona provides sights of the unique colorful rock formations that are present in the area.

On the way home, we followed Bill in our Cherokee, flying behind by several miles. Before we took off, we had agreed on a radio frequency that we could use to communicate with each other. Although the flights from Flagstaff to Byron and Concord were lengthy, we were both able to complete them without having to stop to refuel.

As Dori and I have reviewed our travels within the U. S., there was a section of the country that we had not explored. In particular, we had not been to South Dakota, where the Mt. Rushmore National Monuments are located, nor had we been to Idaho to visit the Lakes of Coeur d'Alene and Pend Oreille. Since the distances between these destinations were large, our Cherokee provided us with the means of covering a lot of ground quickly and efficiently.

Our trip to Coeur d'Alene was made easily in one day with a single fuel stop at the airport in Sun River. As we landed in Coeur d'Alene, it appeared that a weather system was moving in from the west. Our stay in Coeur d'Alene was planned for three or four days, which was good because the approaching weather system we had observed gave us significant rain for the next several days. Although the lake at Coeur d'Alene was very pretty, we actually thought that the Lake Pend Oreille, being somewhat larger and less developed, was more attractive to us.

Another large lake we were interested in seeing was Flathead Lake in Montana. We flew from Coeur d'Alene to Kalispell, Montana, located on the north side of Flathead Lake. In this area we located two couples who were friends of ours from our condominium complex in Mexico. One couple had a house in Kalispell. The other couple had a condominium located near the southern end of Flathead Lake. Here again, we were impressed by the beauty of the lake and by the great boating opportunities that this very large lake seemed to afford.

From Kalispell we flew to Rapid City, South Dakota. From our motel room in Rapid City, we spent several days exploring the area, visiting Mt. Rushmore, and the Dakota Bad Lands. The scenery in the Bad Lands was more dramatic than we had anticipated. In between these more lengthy trips, we flew to Reno a couple of times, to Las Vegas, and several times to Palm Springs.

CHAPTER 59

FOR YEARS I HAD HEARD and read about the big annual air show that takes place at Oshkosh, Wisconsin. One of the flying magazines that I subscribe to advertised that the Piper Owners Society was providing Piper owners an opportunity to land at an airport nearby Oshkosh. From this airport, the Piper Owners Society provided daily transportation to Oshkosh plus evening banquets, and seminars covering maintenance aspects of Piper aircraft.

We signed up for the Piper owners' activities and left from Byron in our Cherokee several days before the Oshkosh show was to begin. The trip to Oshkosh turned out to be straight forward, as we flew non-stop to Ogden, Utah, for a fuel stop, and then on to Scotts Bluff, Nebraska, where we spent the night. The next day was a relatively easy leg, flying from Scotts Bluff to Stevens Point, near OshKosh. When Piper aircraft arrived at Stevens Point, aircraft owners were offered the opportunity to participate in a show and tell for each of their airplanes. Because I had been an early arrival, I was the first owner to do my show and tell.

The Piper Owners' Organization had done a fine job of keeping us entertained and providing the bus transportation that allowed us to avoid the very heavy vehicle traffic in and around the Oshkosh Airport. The spectacle at Oshkosh has to be seen to be believed. Roughly ten thousand aircraft of all types were parked in extended parking areas consisting of some paving but mostly grassy fields. Many of the owners with their families and friends set up camp beneath the wings of their airplanes. Although the conditions were very crowded, everything seemed to be orderly and well-managed.

As we had imagined, the show itself was spectacular. I especially enjoyed the flyovers by the piston-powered warbirds and enjoyed the ground displays of the many different types of aircraft. There were also many large tents where vendors displayed their latest equipment and electronics. All in all, I enjoyed very much the four or five days we spent at Oshkosh and Stevens Point.

As part of the trip, we had selected another destination that we wanted to visit before returning home. There had been much talk and advertising discussing the forms of entertainment available in Branson, Missouri. We departed Stevens Point and flew non-stop to Springfield, Missouri. In Springfield, we rented a car and drove to Branson.

When we arrived in Branson, we were surprised at how crowded the main road through the area was. In fact, the two-lane road through the main area was packed solid with cars and RVs, virtually creeping along. Not having motel reservations, we were nevertheless able to locate a reasonable one not far from the main areas. We stayed in Branson about three nights and saw four major shows. The shows were OK, but not up to Las Vegas standards. On the other hand, they were a lot less expensive too.

Since Springfield, Missouri, is a bit farther south, we selected a more southerly route for our return trip. This choice was a bit of a mistake, since summer monsoonal moisture had begun to come up into the southwest U. S. As we continued to fly west, there was some vertical buildup which resulted in us having hours of rough riding, watching our rate of climb instrument showing a rise of one thousand feet per minute, followed by a sharp downward vector of one thousand feet per minute. We were relieved when we got to the California border and our tiresome bouncing had ceased.

CHAPTER 60

AFTER HAVING BEEN IN OUR condominium complex in Puerto Vallarta for five or six years, some of the good friends who we had enjoyed being with so much had either passed away or moved on. We had also rather thoroughly explored all of the areas within one hundred miles to the north, south, and east of Puerto Vallarta. We had also traveled to many of the colonial cities in the interior.

We, therefore, decided to place our condominium up for sale. We put notices up on all of the bulletin boards within our complex and had made several real estate agents in the area aware of its availability. During the first year, we got virtually no response. The second year that it was advertised some interest had been shown, but no one had been ready to buy.

In January of 2007 a situation had developed where a person in the complex had sold his three-bedroom unit. He had been planning to buy another condominium in a different condominium complex located nearby. This person had had a run in with the sales people at the other complex and decided to back out of buying one of those condos. Thus, he needed a place to live quickly because he needed to move out of his condo by the end of the month. When he was made aware of our place, he looked at it and wanted to buy it, provided that we could move out by the end of January.

Since we sold the condo fully furnished, we agreed that we could leave within the three weeks remaining. Fortunately, it was a cash deal, and because he had a Mexican lawyer who was a friend we were able to make the real estate closing quickly and easily. All we had to do was pack up and remove our personal belongings once we were assured that the deal would go through. Separately, we also sold the rights to the garage that we had leased from the developer. We packed up our Toyota van that we had used in Mexico for the past six years and drove home.

We had some mixed feelings about having sold our beautiful Mexican condo. We had enjoyed many good times with friends that we had made in Puerto Vallarta and the times that we were joined by some of our family. On the other hand, having ownership of this rather expensive property, we felt an obligation to use it and maintain it.

Our use of it meant that we were not doing other things that we might have wanted to do. In all, we had been in Puerto Vallarta over the winters of seven years. As I explained earlier, we had visited all the local areas within a radius of roughly one hundred miles from Puerto Vallarta. All in all we had a sense of relief that the sale of our condo had been accomplished so cleanly and that we were now free to pursue other activities.

CHAPTER 61

AFTER WE ARRIVED HOME, WE decided to make some acquisitions and improvements in our home that we had put off for some time. From time to time we had discussed the possibility of putting a baby grand piano in our living room. We purchased one and Dori began to pursue the playing of the piano, trying to regain the skills she had lost from not having played for so long.

The principle TV that we had in our den was more than twenty-five years old. Although it was still working well, its CRT tube technology had been overtaken. We decided to procure a completely new home entertainment center, which was planned around a fifty-five inch flat-screen TV.

The house that we had built back in 1977 was getting older. Over the years, however, we had always kept the house in a very high state of repair. Various functional components that tend to wear and degrade we replaced and repaired long before they failed. The house was kept up to date by redoing the kitchen and retiling the entire first floor. Other major improvements and upgrades included rebuilding the dock and both the lower and upper decks attached to the back of the house. We also replaced the fence on the side of the house by the front door with a stucco wall. The wall created a courtyard which is landscaped with lush greenery.

CHAPTER 62

IN THE VERY BEGINNING OF our life in Discovery Bay, we were part of a small group that founded the Discovery Bay Yacht Club. Being a member of this club enhanced the use of the several different boats that we have owned. In more recent years, it has been traditional that the club sponsors several week-long cruises to the San Francisco Bay, as well as many weekend cruises during the year. We have been, and continue to be, participants in these cruises.

We typically spend three or four weeks in the bay during September and October. Along with other cruises we participate in the delta region, we have found that our recent boat usage totals between forty to fifty nights that are spent aboard the boat.

About seven or eight years ago some of our boating friends decided to join Weber Point Yacht Club, a small yacht club, located just off the San Joaquin River, near Stockton. This small club, which is only accessible by water, has only thirty-six memberships, as opposed to the nearly five hundred memberships in the Discovery Bay Yacht Club. For several years we delayed joining this club because of our commitments to the property we still held in Mexico.

About four years ago we decided to join and have become active participants.

Most of the repairs and improvements to the club are performed by the membership of WPYC. Major scheduled events take place monthly, during the yachting season. Two or three couples get together and organize each event and provide food and entertainment for the membership. Each year we have been volunteering to be one of the hosts of an event. In addition to these routine Weber Point activities, Dori and I have taken on other responsibilities. Dori is editor of The Spyglass, which is a newsletter for WPYC, which is published five or six times a year. As a result of taking on this responsibility, she has had to significantly upgrade her computer skills. I have become one of four directors who provide advice to the flag officers of the club. In addition to the usual director responsibilities, I am also the club's safety officer. In this capacity, I keep track of safety issues and make recommendations for safety improvements or changes to the governing board.

CHAPTER 63

IN THE WEBER POINT YACHT Club, there are two other active pilots. In conversations that I had with one of them, we discussed the possibility of taking a long flight together. The idea for such a trip developed as a result of two other members who were touring the southeast in their motor homes. We thought that it might be fun to join up with them somewhere along the way. Part of the motor home schedule was that our friends planned to be in New Orleans for about a week. New Orleans became the destination of the flying trip that we would take.

The other pilot, Nels, and his wife, Judy, had a more capable airplane than our Cherokee. Their plane was a six-place retractable landing gear Bonanza B36. It had an autopilot with altitude hold, in addition to heading hold which was all my autopilot had. The Bonanza was also turbocharged and had a built in oxygen system, which meant that we could fly higher, if necessary.

Due to these advantages that the Bonanza offered, we decided to use it rather than my Cherokee. Nels and I shared flight-planning work. Nels prepared a spread sheet that summarized all the information we might need more completely than anything I was accustomed to preparing for myself.

It was surprising that, with the four of us and a total of about sixty pounds of baggage, the six-passenger plane could not fly with full fuel tanks without being over gross weight. We could only carry eighty gallons of fuel, rather than the full capacity of one hundred and two gallons.

As we flew toward the east, the plane performed well. Most of the time we flew at an altitude of between fourteen and sixteen thousand feet. Because our oxygen supply was somewhat limited, Judy and Dori had to go without oxygen much of the time. They did lots of sleeping. The entire trip to the east, while bumpy sometimes, was uneventful.

When we arrived in New Orleans, we were met and picked up by our friends, Dick and Vernie and Rich and Lee. The motorhome couples had something planned for us every day. We took a tour of the bayous in a pontoon boat. There was lots of wild life to be seen, including good-sized crocodiles swimming in the water. To liven up the group, a baby crocodile was passed around for everyone to hold. Fortunately, its mouth was taped closed. We toured some of the old plantations that dated back to Civil War days. The plantations themselves and the surrounding grounds and gardens were beautiful.

We also took a tour of the Ninth Ward. The Ninth Ward consisted of an area where the most severe flood damage from Hurricane Katrina had occurred. On another day, we toured the Marti Gras Museum and storage area. The large, elaborate floats are changed superficially each year, but they use the same basic platform over and over. One of the floats that we viewed carried as many as forty participants. It even had rest-room facilities on board.

We spent one afternoon walking along the Mississippi River front. The river front in New Orleans is heavily developed, having an assortment of restaurants and even some large shopping centers. Another day we walked the length of Bourbon Street. The sights and sounds that we witnessed on Bourbon Street were even beyond any expectations of wildness that we may have had. Part of our explorations included finding many good places to eat. One special restaurant, noted for its breakfasts, gave you enough food to last the entire day.

After the great times that we had in New Orleans, we loaded up the Bonanza and headed home. Since the autopilot was used almost exclusively while we were in cruise flight, I did practically none of the hands-on flying. I did, however, handle most of the navigation and radio communication chores. This took a fair amount of the burden off of Nels.

One of our intermediate stops for fuel on the way home was in Flagstaff, Arizona. At Flagstaff, the fueling station was about a quarter to a half a mile away from the main terminal. As we deplaned at the fueling station, Nels and I took care of the refueling process, while Judy and Dori headed to the main terminal in order to use the rest-room facilities and order lunch.

As Judy and Dori approached the main terminal, a cart came chasing after them. The driver belligerently asked them where they were going. When they replied that they were looking for rest-room facilities. The driver said, "Do you realize that if you had gone another twenty feet you would be charged a ten thousand dollar fine each for violating a secure area?" There were no warning signs that indicated that the area was restricted. Furthermore, the red line painted on the concrete had deteriorated so that it was barely visible. Our wives were rather upset by the whole episode. Because of the possibility of such severe penalties, more obvious warnings need to be provided.

Because of the lengthy flights, much of them being flown at high altitude, it took a couple of days after we arrived home to rest up a feel back to normal. In all, it was a memorable trip and experience.

CHAPTER 64

ABOUT FIVE YEARS AGO I became involved in a project that was tasked to look into how much and what type of information regarding the design of nuclear weapons exists in the unclassified domain. The concern over the amount and accuracy of the information is due to the potential this information could have toward further proliferation, and to how it could lead to terrorists having the capability to build an improvised nuclear device (IND).

After development of a set of unclassified search terms, a computer search engine is employed to search for sources on the internet and in books and publications containing any of the search terms. The amount of the material that has been collected is large. Some of the largest accumulations occur in a relatively few web sites and publications since there are a few knowledgeable, smart individuals who have essentially made a career of finding and publishing everything that they can find having to do with nuclear weapons.

As the compilation of weapons related material is brought into the classified domain (into a secured area), we narrow the search further by applying classified search terms. If any of the classified search terms appear, the finding is subjected to further analysis.

I am one of several senior people with actual weapon-design experience who assess the accuracy and the significance of the material found. From this work, the group has completed a number of reports for the DOE headquarters. These reports each cover a particular category of weapon design. Examples of these categories include gun assembled weapons, implosion weapons, and thermonuclear weapons.

We understand that the reports completed have been well-received in Washington. As a result, our sponsorship and funding has been maintained. My involvement in this project is continuing.

CHAPTER 65

I HAVE NOT COVERED THE many cruise line cruises, river boat cruises, and other trips we have taken. Such traveling does not for the most part involve significant interaction from us in their execution. Such traveling, however, has taken us all over the globe, including two trips that took us completely around the world. We recently counted 59 countries and significant places that we have visited. A listing of these countries and places is given below:

Britain
Scotland
Wales
Ireland
France
Belgium
Holland
Germany
Denmark
Sweden
Finland
Russia
Italy
Greece
Albania
Bulgaria
Turkey
Israel
Jordan
Egypt
Oman
UAE
Dubai
India
Korea
China
Australia
Canada
Mexico
Guatemala
Honduras
Panama
Venezuela
Brazil
Uraguay
Argentina

Poland	Thailand	Falkland Is.
Czech Republic	Singapore	Chile
Austria	Japan	Spain
Gibraltar	Monaco	Luxemburg
Bahrain	Croatia	Montenegro
Kwajalein	Christmas Island	
Puerto Rico	San Martin	
Virgin Islands – American and British		
Barbados	Bahamas	

Obviously there are numerous memorable happenings that occurred on these trips. There are three trips that stand out most vividly in my memory. One, of course, was one taken in June of 2004 in celebration of our 50th Wedding Anniversary. We took our entire family (14 of us) on a Princess cruise to Alaska. We had four cabins, including a mini-suite for us. The mini-suite was used by all of us to serve as a meeting point to get together and enjoy evening happy hours. Our grandchildren were old enough that they could enjoy the freedom of being able to move around the ship on their own, but still young enough that they were happy being with their parents and grandparents. All in all, the trip turned out to be a tremendous success and was a great way to have made our anniversary a cherished memorable event.

Another trip involved a cruise through the Greek Islands. On this trip, we spent about a week basing ourselves in Athens while we did some land touring in a rented car. Following the land touring, we boarded a smaller Greek cruise ship that visited the Greek Islands, the west coast of Turkey, the Bosporus, and Istanbul.

For the land portion of the trip we chose to stay in a Marriott Hotel a bit outside of the center of Athens. I was seriously into running at this time. I had noticed that the Parthenon was only two or three miles from our hotel. While staying there, I initiated a routine where I would awaken before dawn, don my running clothes, and run to the base of the Acropolis. From there, I ran up the Acropolis to the Parthenon and waited for the sun to rise.

As I stood there surrounded by the beautiful ancient structure, I was awed by the spectacle as the sun began to light up my surroundings. Those daily morning runs were clearly the most impressive of any that I have ever experienced.

The third outstanding trip that I have taken was one to Christmas Island. In the fall of 1961, the Soviets broke the comprehensive nuclear test ban treaty that both the Soviets and the U. S. had been honoring since 1958. The Russian atmospheric tests were carried out at a rapid rate. Their series included some at very high yield. In fact, one of their tests had a yield far higher (~50 MT) than any test ever conducted by the U. S. It was clear that the Russian test series had been planned for some time.

The breaking of the treaty caught the U. S. by surprise. The U. S. followed suit as soon as possible, by conducting its own series of high-yield tests at Christmas Island (owned by the British). As part of our test series, some people involved with nuclear design work were invited to take tours at Christmas Island in order to witness the test activities and a full-scale nuclear test, if it occurred while you were there. I was selected to take one of these tours.

We first flew to Honolulu, Hawaii, staying overnight in the Royal Hawaiian Hotel. The following day we boarded a military aircraft for our trip to Christmas Island. We arrived in the evening and were housed temporarily in some old British barracks that may have been there since World War II. The next day we were moved to our quarters, which consisted of a series of large tents with wooden floors.

The next day several of us were given a jeep with which to explore the island. We were warned to be careful about the intensity of the sun because Christmas Island is very close to the equator. Many of the roads were unimproved. Some were just trails in the sand. As we drove through the sandy trails, we got our jeep stuck. We pushed and shoved for an hour or more before finally getting free. It was hot so I was dressed only in shorts and shoes. With no sun protection, I got terrible sunburn. I could hardly get off my cot as the pain and throbbing in my swollen, burned, legs made it an agonizing effort to stand up.

After several days, I got better and looked forward to witnessing the test of a high-yield thermonuclear device that was to take place the next day. The planned yield for the device to be tested was eight megatons. The device was to be loaded into a B52 bomber at Barber's Point on the Island of Oahu. From there, it would be flown to Christmas Island and dropped. The planned drop would be from about thirty thousand feet, with the bomb fusing set for detonation at eight thousand feet. The distance to the island beach from the detonation point was to be twenty-two miles.

On the morning of the planned test, we were awakened while it was still dark. As we left our tents, we wrapped ourselves in white sheets to protect ourselves from the thermal effects of the explosion. Everyone was given a set of protective goggles, which were so dark that in normal daylight practically nothing could be seen. While it was still dark, the count-down began, as the B52 approached its release point. Finally, a voice on the loud speaker announced that the bomb had been released and that we should be wearing our goggles and that we should turn away.

Suddenly, even with the dark goggles, everything lit up brighter than on a sunny day. As we turned around to observe the fireball, it had grown to an enormous size. You could feel the intense heat on your face. So far, there was no sound. Because of the size of the fireball, I was expecting a tremendous noise when the blast wave arrived. As we stood there watching the beautiful colors developing in the fireball, a wind came from behind us. The wind signaled that the blast wave would arrive shortly. To my surprise, the blast and sound was not much greater than that produced by the firing of a large cannon.

By this time, it appeared that the fireball light had diminished enough that we could remove our goggles. As I began to pull them away from my face, the light was still blindingly intense. The goggles needed to stay on for several more minutes. The beauty of the colors in the fireball and the enormity of the display was beyond anything that I could have expected. It truly was one of the most memorable events of my life.

Since atmospheric testing ceased in 1962, there are now few surviving people who have ever had the experience of witnessing an atmospheric nuclear explosion, especially one as large as the one I was privileged to have observed.

Flight to Guadalajara in Piper – 2002

BILL AND DORI SCANLIN 50 YEARS

Bill and Dori Scanlin of Discovery Bay are celebrating their 50th anniversary on a cruise to Alaska with their children and grandchildren.

Both Bill and Dori were born in Philadelphia, Pennsylvania. They were married June 26, 1954 and lived in northern New Jersey during the first years of their marriage. They moved to California in 1960. Bill worked for Lawrence Livermore National Laboratory for more than 30 years. Dori worked for the Livermore Unified School District for 20 years. Both retired in 1991. When Bill retired, he was Deputy Associate Director for the Livermore Nuclear Weapons Program.

Their three children and their spouses and six grandchildren will accompany them on the cruise. They are Susan and Richard Klekman, and their children Jaclyn and Matthew; Linda and Wayne Fulton, and their children Robert and Kenneth, and Bill and Alissa Scanlin and their children Timothy and Christopher.

Bill and Dori are avid boaters and active in the Discovery Bay Yacht Club. Bill still owns his own plane which they frequently use to visit their children and travel. They are both active members of the Discovery Bay Health Club.

50th Anniversary in 2004

Alaskan Cruise with family in 2004

CHAPTER 66

OVER THE YEARS, WE HAVE traveled to so many places in the world that our most recent travels have tried to focus on going to places we haven't yet been. Consequently, in January 2010, we took a cruise around South America, starting in Rio de Janeiro and ending in Valparaiso, Chile. We were lucky in our timing because Valparaiso and nearby Santiago were significantly damaged by the recent Chilean earthquake just weeks after we left. In May of 2009, we took a driving trip around Eastern Europe. As a result of the previous communist occupation, we had long-avoided travel in some of that area.

While we are able to maintain our health and motivation to travel, Dori and I plan to continue traveling to unexplored destinations and to visit again some favorite places. There are still many places in the U. S. that we would like to visit. For more distant destinations, we prefer using commercial air and car rental. For traveling to destinations closer to home, our Piper Cherokee and rental cars remain an option.

One item that I have not covered has to do with our most recent vehicle acquisition. I am mentioning this because the vehicle that we purchased is so special. Several years ago, when we returned from Mexico with our Toyota van, we realized that the van had little resale value, even though it was in great shape. We decided to keep it. However we still had the Ford F350 truck we had used to tow the Sea Ray we had back east.

The truck looked brand new and had only 12,000 miles on the odometer. The truck was no longer needed to haul bulky items since we had the van. We decided that due to the truck's nearly new condition, we should be able to get a good price for it. We took it to a car show at the Pleasanton Fairgrounds. We got only one serious buyer and he bought it. The price we got was about half the amount that we had paid for it fifteen years before.

The garage stall where the truck had been was now empty. In looking at car ads one day, we noticed that a nearby Chevrolet dealer was selling Corvettes for $7,000 below retail price. For years, I had always longed for a great sports car like a Corvette, but my practical and somewhat frugal side kept me from pursuing such a luxury. In the case of these Corvettes, I decided that this might be an opportunity to satisfy my long-standing desire.

As Dori and I shopped, we were taken by the great looks of the Corvette Z06 model. The specs on the Z06 version are incredible. Overall, it is quite a different car than the standard model. Its handling, acceleration, and speed capability is beyond anything else you can buy without spending in the vicinity of $200,000. After a test drive I was hooked.

We bought a silver Z06 in April 2008. The Z06 is now our special occasion car. If it is raining or road conditions may not be good, I won't take it out. Even though it is now two years old, it looks as good today as the day I brought it home.

I have written most of this book during the winter of 2010. As we have for the last three years, we have spent two months in a rented condominium in Palm Desert, California. The weather in this area is near perfect this time of year. The nights are cool and the days are warm. We plan to be here again next year, enjoying some of our (mostly boating) friends from Discovery Bay who also winter here.

Most people who read this book would likely consider that I have lived a dream life. For me it certainly has been. Most importantly, I am happily married, as I have been for fifty-six years, to a wonderful, lovely wife. Also, I am blessed to have loving children and grandchildren that I am continually proud of. In addition, it has been very important that Dori and I continue to enjoy good health.

At work, I had job assignments that were important, tremendously satisfying, and at all times challenging. Overall, the mental and physical accomplishments I managed to achieve were beyond anything I could have expected or imagined as a young man. I am also fortunate to have a wife who enjoys adventure as much as I do. The episodes I have related in this book are testimony that we certainly enjoy adventure. I am hopeful that the great life we have had can continue to be sustained in the coming years.

The boys – Chris, Robbie, Tim, Kenny, and Matt
Our Grandsons

The women – Alissa, Sue, Jackie, Dori, and Linda

Hearst Castle in 2007

Latest Toy – 2008